Walter Tromm (Ed.)

Annual Report 2020 of the Institute for Thermal Energy Technology and Safety

Karlsruhe Institute of Technology
KIT SCIENTIFIC REPORTS 7762

Annual Report 2020 of the Institute for Thermal Energy Technology and Safety

Edited by
Walter Tromm

KIT Scientific Publishing

Report-Nr. KIT-SR 7762

Impressum

KIT Scientific Publishing

Karlsruher Institut für Technologie (KIT)
KIT Scientific Publishing
Straße am Forum 2
D-76131 Karlsruhe

KIT Scientific Publishing is a registered trademark
of Karlsruhe Institute of Technology.
Reprint using the book cover is not allowed.

www.ksp.kit.edu

Print on Demand 2022 – Gedruckt auf FSC-zertifiziertem Papier

ISSN 1869-9669
ISBN 978-3-7315-1153-3
DOI 10.5445/KSP/1000140139

Content

Institute for Thermal Energy Technology and Safety ... 1

Structure and Activities of the Institute for Thermal Energy Technology and Safety (ITES) ... 1

Resources ... 1

Group: Magnetohydrodynamic ... 5

Magnetohydrodynamics for liquid-metal blankets ... 5

Introduction ... 5

Numerical simulations for liquid metal MHD flows ... 5

MHD velocity distribution and pressure drop in manifolds of a WCLL TBM ... 8

Design and fabrication of a WCLL TBM mock-up for MHD experiments ... 11

Reconstruction of 3D velocity profiles from measured surface potential data ... 12

Experimental study of heat transfer in magneto-convective flows related to WCLL blankets ... 15

Further work ... 17

References ... 18

Group: Severe Accident Research ... 21

SAR/ITES research activities in 2020 ... 21

Group: Multiphase Flow ... 25

Detailed investigations on flow boiling of water up to the critical heat flux ... 25

Introduction ... 25

COSMOS-L ... 25

COSMOS-H ... 28

Acknowledgements ... 30

References ... 30

Group: Karlsruhe Liquid metal LAboratory (KALLA) ... 31

Liquid metal technology research at Karlsruhe Liquid Metal Laboratory 2020 – Fueling sustainable energy and process technology by innovative approaches ... 31

Introduction ... 31

Thermal Fluid Dynamics of Liquid Metals ... 32

Liquid Metal Based Heat Storage ... 34

Liquid Metal Based Process Technology ... 35

Group: Accident Management Systems ... 39

Smart Resilience Engineering: Mitigative Planning of Networked Infrastructures, Uncertainties and Robust Management ... 39

Introduction ... 39

Resilient Smart Grid Design and Microgrids ... 40

Resilient Energy Management ... 41

Dynamic Measure Development – Agent-based optimization ... 44

Nuclear Preparedness and Response 44
Ensemble Modelling and Uncertainty 45
References 47
Group: Hydrogen 51
Fundamental Behaviour of Hydrogen to Applied Accident Consequence Analysis for Hydrogen as a Safe Energy Carrier 51
Introduction 51
Liqud Hydrogen Pool Experiments 51
Transient cryogenic jet ignition 54
CryoTube Experiments 55
References 57
Group: Energy and Process Engineering 59
First experimental results of a supercritical Organic Rankine cycle using propane as working fluid 59
The MoNiKa power plant 59
Components description 59
Instrumentation 61
Operational design parameter 62
Simulation Software GESI 62
Test run 63
Summary 65
References 65
Group: Framatome Professional School 67
Annual report of the Framatome Professional School 67
CFD Validation of heated rod bundles in heavy liquid metals 67
Model validation 68
Creation of a reduced order model for the k - ε model 70
A free programmable lab-on-a-chip 71
BattMarines 73
List of Publications 75

Structure and Activities of the Institute for Thermal Energy Technology and Safety (ITES)

Walter Tromm

The Institute for Thermal Energy and Safety (ITES) is situated with its offices and research laboratories in Campus North of Karlsruhe Institute of Technology (KIT).

The research of ITES focuses on conversion from thermal power to electric power for future power systems without CO2 emission including hydrogen technologies and on safety. In the past ITES concentrated on safety features and on methods to mitigate severe accidents for nuclear power plants. Nowadays, ITES uses these competences, gained from a profound experience in numerical simulation and in design and operation of large-scale experimental facilities to apply this knowledge mainly in the area of renewable energies.

The combination of science and technology with education and training is a systematic approach at KIT, and ITES is contributing accordingly to courses in mechanical engineering, supervises several bachelor and master theses each year and coordinates master programs in energy technologies. Compact courses on energy technologies are given also in executive master programs and in the Framatome Professional School, which is funded by industry and managed by ITES.

Resources

The employees of ITES contribute to the research-programs Renewable Energies (EE),

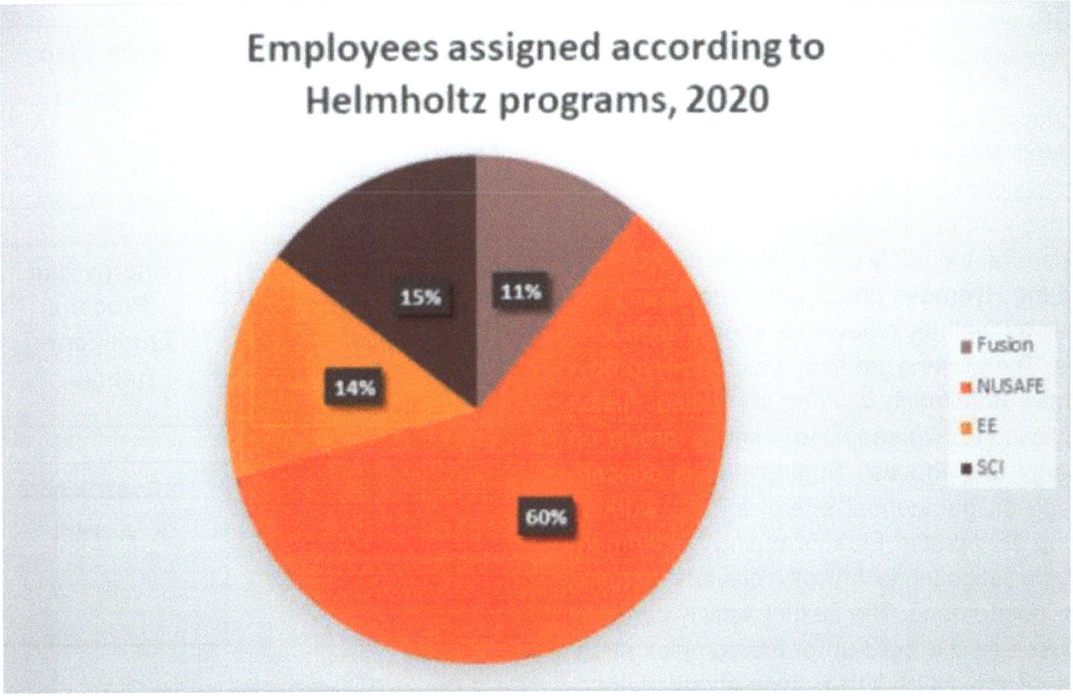

Fig. 1 Assignment of ITES personnel to helmholtz-programms

Storage and Cross-linked Infrastructures (SCI), Fusion (FUSION), and Nuclear Waste Management, Safety and Radiation Research (NUSAFE) in the research field energy of the Helmholtz Association (HGF). By the end of 2020, around 80 scientists, engineers and technicians have been working at ITES on this wide range of CO_2-free technologies for energy conversion. Fig. 1 illustrates that still about 60 % of the ITES employees were working for the NUSAFE program, but steadily reduced in recent years. Around two third of the employees were funded in 2020 by the Helmholtz Association (HGF), the others by third party funds of the European Commission, by industry, by German ministries or by other research funds. Doctoral students as well as students of the Baden-Wuerttemberg Cooperative State University (DHBW) were filling around 20% of these positions at ITES. In addition, students perform their bachelor or master theses or spend an internship in the research laboratories of ITES.

An overview of the structure of ITES is given by the organization chart, Fig. 2. Because all working groups are embedded in the research field energy of the Helmholtz association (HGF) a close collaboration within the groups is guaranteed.

Working groups on Severe Accident Research, on Multi-Phase Flow as well as the Framatome Professional School still concentrated in 2020 primarily on nuclear applications, whereas the Karlsruhe Liquid Metal Laboratory (KALLA) and the Hydrogen group were addressing nuclear as well as renewable energy technologies. The working group on Magneto-Hydrodynamics is primarily working on nuclear fusion applications, whereas the working group on Energy and Process Engineering is concentrating on geothermal energies. The working group Multiphase Flow is mainly experimentally investigating technically relevant, complex flow phenomena. The field of activity currently focuses on the build-up of the complex structure of the COSMOS-H high-pressure loop. The working group on Accident Management Systems continued to extend the application of its simulation models to several critical infrastructures. Thus, the institute covers a wide field of different energy technologies and related safety investigations, and the share of personnel resources on the different research topics reflects the requirements of the Helmholtz Association's energy research field.

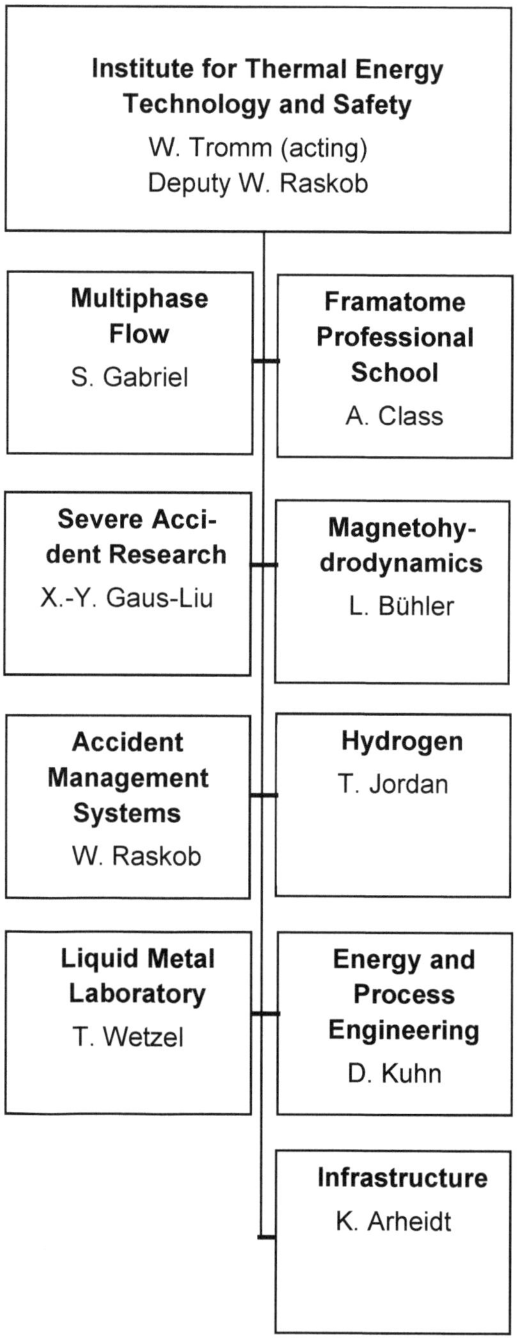

Fig. 2: Organization chart of the Institute for Thermal Energy Technology and Safety

They are all supported by a joint infrastructure, comprising a metal workshop, manufacturing urgent test components, a welding shop, and an electromechanical workshop. Other tasks of the infrastructure include the IT-administration, business administration and public website of the ITES. The Infrastructure team is active as well in education and training activities.

Every year, at least 4 students of the Baden-Württemberg Cooperative State University are employed by ITES, managed by the Infrastructure group, to work with the research teams as part of their educational program.

Working in a research area with industrial application ITES practices a Quality Management System appropriate to EN ISO 9001.

Group: Magnetohydrodynamic

Magnetohydrodynamics for liquid-metal blankets

L. Bühler, H.-J. Brinkmann, C. Courtessole, V. Klüber, C. Köhly, C. Mistrangelo

Introduction

The design of breeding blankets represents a major challenge for fusion reactor engineering due to performance requirements and severe operating conditions in terms of heat load and neutron flux. Liquid metal alloys such as lead lithium PbLi are considered as breeder material due to their lithium content, as neutron multiplier, and neutron shield. Because of their large thermal conductivity and the possibility to be operated at high temperature, liquid metals may serve additionally as coolants to remove the nuclear heat released in fusion reactors. On the other hand, when the electrically conducting fluid moves in the strong magnetic field that confines the plasma, electric currents are induced. The latter are responsible for strong electromagnetic Lorentz forces and high pressure drop. Magnetohydrodynamic (MHD) pressure drops have to be carefully quantified, since excessive values can jeopardize the feasibility of the considered blanket concept. Moreover, Lorentz forces affect the velocity distribution and suppress turbulence with implications on heat and mass transfer.

The physical parameters characterizing the MHD flow are the Hartmann number, the Reynolds number and the Grashof number

$$Ha = BL\sqrt{\frac{\sigma}{\rho\nu}}, \quad Re = \frac{u_0 L}{\nu}, \quad Gr = \frac{g\beta\Delta T L^3}{\nu^2},$$

where B, L and u_0 denote magnetic field, typical length scale and characteristic velocity. The fluid properties are density ρ, kinematic viscosity ν, electric conductivity σ and volumetric thermal expansion coefficient β.

Ha^2 and Re stand for ratios of electromagnetic to viscous forces and inertia to viscous forces, respectively. The importance of buoyancy forces is quantified by Gr, where ΔT is a typical temperature difference and g stands for gravity.

The MHD work at ITES KIT supports DEMO blanket engineering activities as part of the EUROfusion consortium and the test blanket program of Fusion For Energy (F4E) for ITER.

Numerical simulations for liquid metal MHD flows

Simulation of 3D MHD flows in pipes with flow channel inserts

When duct walls are thick and electrically conducting, flow-induced currents find shortcuts along these walls, resulting in high current density in the fluid, strong electromagnetic forces and considerable pressure drop. In order to mitigate MHD effects in fusion blankets, electrically insulating flow channel inserts (FCI) have been proposed. They are loosely fitted in liquid metal ducts or pipes with the purpose of electrically decoupling the fluid region from the well-conducting blanket structure. [1]. In the following, a FCI of sandwich-type is considered, as schematically shown in Figure 1, which consists of an insulation layer (orange), enclosed by two thin sheets of steel to protect the insulation from fluid infiltration. Of particular interest are 3D MHD effects present when the flow enters or exits the FCI or at gaps between two FCIs. At these locations current density is locally increased by extra 3D currents j_{3D}, which affect the flow. This leads to local modifications of electric potential, pressure and velocity. Numerical investigations of complex FCI entry flows at fusion relevant parameters $Ha = 2000$, $Re = 20000$, allows to quantify these 3D effects and provide insight into related experiments performed in the MEKKA laboratory at KIT.

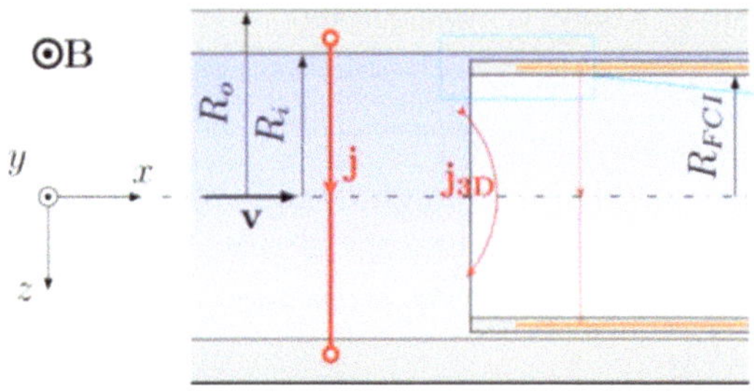

Figure 1 Sketch of the MHD flow entering a FCI, as tested for a circular pipe in the MEKKA laboratory. The orange surface indicates the insulation layer.

The computational model considers the FCI entrance flow at a high degree of geometrical detail, which allows the resolution of wall and conducting sheets of the FCI for determination of currents. The geometry, which takes into account the liquid metal filled gap between FCI and pipe wall, is shown in Figure 1.

In Figure 2 the electric potential distribution is shown in the lower half of the pipe. The incoming flow is fully developed far upstream with no variations of potential along x and along magnetic field lines. Here, currents close in cross sectional planes, as shown by current lines displayed in the upper half of the geometry at $x=x_1$, x_2. Potential and currents in the other halves of the pipe behave in a y- symmetric way and they are in agreement with known solutions for fully developed MHD flow under a uniform magnetic field (e.g. [2]).

After entering the FCI for $x>0$, the insulation prevents fluid currents from entering the thick-wall domain. They are blocked by the insulation and forced to close along the thin inner sheet of the FCI (red current lines). Far enough downstream, the flow approaches again fully developed conditions.

Three dimensional effects near the FCI entrance at $x=0$ lead to axial components of current density in the fluid and in the wall. This is caused on one hand by the fact that the potential in the outer wall drops to zero for $x>>1$, since no currents are supplied to this part of the pipe due to the presence of the insulating layer. The resulting axial potential gradient drives 3D currents in the outer wall, as shown by the strongly bent blue current lines near $x=0$. On the other hand, current loops closing along the thin conducting sheet of the FCI experience higher resistance and therefore the potential difference in the cross section is higher than in the well conducting part of the pipe for $x<0$. The reason is the significant difference of wall conductivity. 3D effects on currents and potential persist downstream into the FCI over multiple characteristic lengths.

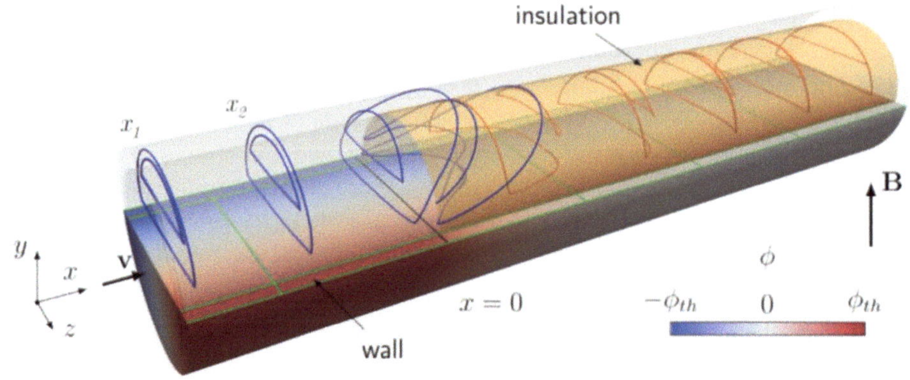

Figure 2 MHD flow entering a FCI for $Ha=2000$, $Re=20000$. Electric potential distribution in the lower half of the pipe and electric current streamlines in the upper half. Current streamlines closing over the wall domain are displayed in blue, while those closing along the FCI are drawn in red. The orange surface represents the insulation layer and the black straight line at $x=0$ marks its beginning.

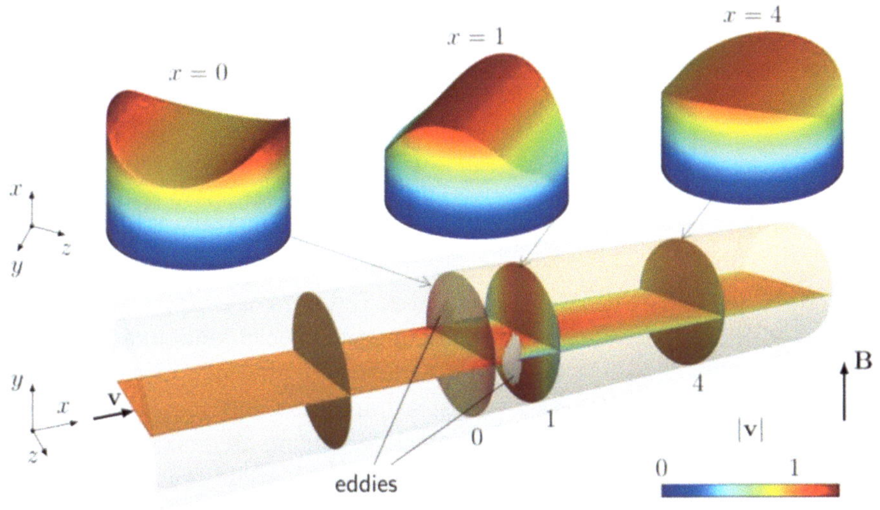

Figure 3 3D profiles of velocity at three axial positions near the entrance of the FCI. The insulation starts at x=0.

Such complex currents interact with the magnetic field, resulting in Lorentz forces that affect the flow. Figure 3 shows the resulting velocity distribution across the FCI entry region at three axial positions downstream in the FCI. The incoming flow, not shown in the figure, has a uniform core velocity and thin boundary layers along the pipe wall.

As soon as the fluid enters the insulated part (x=0), it moves towards the sides forming a concave velocity profile along the transverse z-direction. This rapidly changes into a convex profile with a center peak (x=1). Then, over several characteristic lengths, the velocity slowly converges towards a fully developed flow in the FCI, where the core velocity is almost constant but slightly curved along z, as displayed at x=4. Due to the strong velocity gradients in the entrance regions, side layers may become unstable exhibiting locally confined transient eddy regimes between $0 < x < 1$.

The pressure distribution in the pipe and FCI is shown in Figure 4 along the axis (blue) and near the sides (red). Pressure gradients in MHD flows in bare pipes and in FCI, as predicted by an asymptotic theory, are marked as straight black lines in the plot. Results from 3D numerical simulations perfectly agree with these theoretical predictions, the vicinity of the FCI entry, center and side pressure separate.

While the transition of the core pressure appears relatively smooth, the pressure near the sides experiences a sharp drop in front of the FCI. Extremely high current densities entering the well conducting wall in front of the FCI result in large Lorentz forces, leading to such high-gradient peaks near the sides. After entering the FCI, the side pressure remains almost constant until leveling off with the core pressure. For ideal fully developed conditions in very long pipes, the relatively low wall conductance of the FCI in comparison to that of the bare pipe wall leads to a significant pressure gradient reduction factor of about 16. 3D MHD phenomena occurring at entrance and exit of FCIs may reduce the efficiency of the insulation due to the additional pressure drop Δp_{3D}, caused by recirculating 3D currents.

While experimental results yield data often extracted from measurements on the outer surface of the pipe [3], the presented numerical investigations provide valuable data and insight into 3D MHD phenomena in the fluid. Computations confirm the FCI's efficiency and show good agreement with asymptotic theory and experimental results.

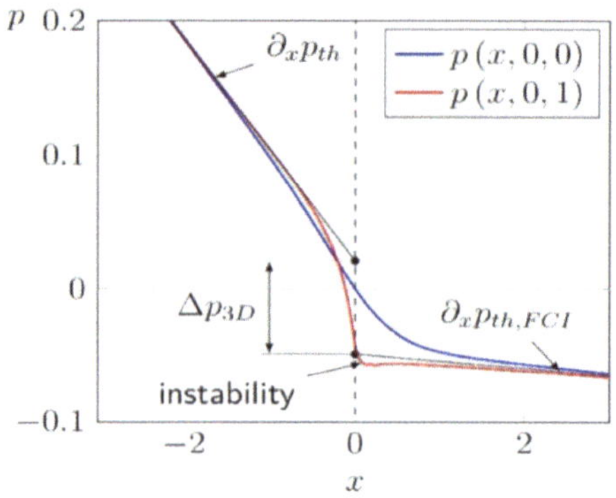

Figure 4 Pressure along the pipe axis (blue) and at the sides (red).

MHD velocity distribution and pressure drop in manifolds of a WCLL TBM

In the frame of the EUROfusion breeding blanket research activities, the water-cooled lead lithium (WCLL) blanket has been selected as "driver" for a DEMO reactor [4]. Test blanket modules (TBMs) derived from this concept will be tested in the International Thermonuclear Experimental Reactor ITER.

In the present study, liquid metal MHD flows in manifolds of a WCLL TBM have been investigated numerically for different sets of flow parameters.

A view of the WCLL TBM manifold design is shown in Figure 5a. It consists of two long poloidal ducts, electrically connected across a common wall. The liquid metal flows along a series of expansions and contractions due to the presence of horizontal stiffening plates that separate the breeder chambers arranged in a column. This type of MHD flow causes additional pressure drop compared to flow in straight ducts [5]. Liquid metal flows are investigated in the model geometry depicted in Figure 5b, whose dimensions are taken from the most recent WCLL TBM design. A uniform magnetic field is imposed in toroidal direction.

Numerical simulations are performed for periodic fractions of the manifolds by assuming a stepwise decrease/increase of the flow in the feeding/draining manifolds. Velocity and pressure distributions in the entire manifolds that feed/drain 8 breeder zones are then reconstructed. For all investigated parameters a uniform flow partitioning among BUs (Breeding Unit) has been assumed and flow rates in the manifolds have been derived by taking the same given mean velocity u_0 in all breeding zones. The pressure drop depends linearly on the mean velocity and quadratically on the magnetic field strength. Therefore, results can be extrapolated and scaled, e.g. to other flow rates, depending on design specifications. For the discussion of the results, we consider the manifold as subdivided into eight parts, each corresponding to a specific fed/drained breeding zone (BZ). For that reason the different portions of the manifolds along the poloidal direction are referred to as BZ1,..., BZ8, as indicated in Figure 5a.

Calculations have been carried out by using a finite volume code. Equations describing the MHD flow are implemented in the open source software OpenFoam [6]. Accurate simulations of MHD flows are pretty demanding since they require a proper resolution of thin boundary layers that form along all walls. Their thickness reduces by increasing the intensity of the magnetic field [7]. Moreover, electrically conducting walls provide closing paths for electric currents and their resolution is crucial to determine the total current density in the fluid. About $4\cdot 10^6$ nodes are needed in the two fluid domains and $6\text{-}7\cdot 10^6$ in the thick wall to resolve an eighth fraction of the manifold. An example of the computational mesh used for the simulations is depicted in Figure 5c in a toroidal poloidal plane. Simulations have been performed on the HPC MARCONI in Italy by using up to 128 CPUs.

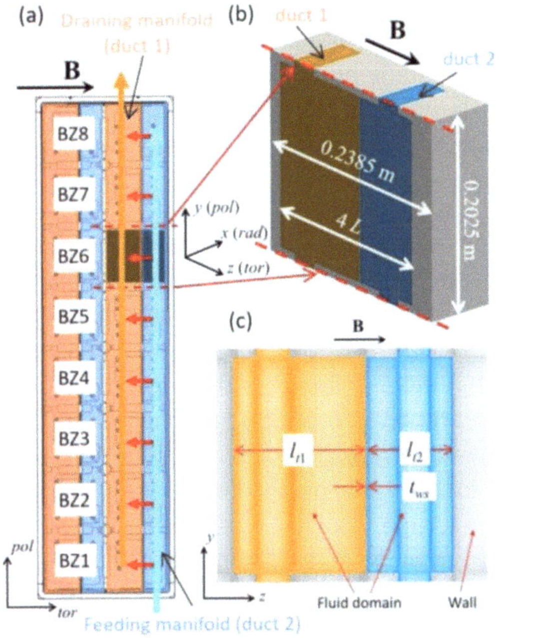

Figure 5 (a) Design of WCLL TBM manifold. Feeding and collecting manifolds are referred to as duct 2 and duct 1, respectively. (b) Periodic fraction of model geometry for numerical analysis. (c) Computational mesh for fluid and wall.

Figure 6 (top) shows, as examples for the flow at Ha=4000 and u_0=0.1 mm/s, 3D views of the velocity distribution at three poloidal locations in the considered geometry at the bottom of the module in BZ1 (a) and in the middle in BZ5 (b). In BZ1 the fluid is supplied to duct 2 (feeding manifold), while the weak flow in duct 1 (collecting manifold) is only driven by electromagnetic coupling caused by leakage currents. In the lower part of Figure 6 the vertical component of the velocity is plotted along the radial white dashed lines in the two channels, as marked in the 3D views. Due to electromagnetic coupling across the common dividing wall the flow in duct 2 pulls that in the core of duct 1 in the same direction. This leads to a buildup of pressure along the flow direction that drives the backward oriented jets in duct 1 (orange curve in Figure 6a at the bottom). In Figure 6b results in the middle portion of the manifolds, BZ5, are displayed. Here the mean velocities in the two coupled ducts are very similar. As a result the velocity distribution is analogous in both channels. Electric currents induced in the cores of the manifolds enter the corresponding side walls, they flow towards the external and common Hartmann walls through which they pass into the opposite parallel wall. Consequently, no current is exchanged between the two ducts and no significant multichannel effects are observable.

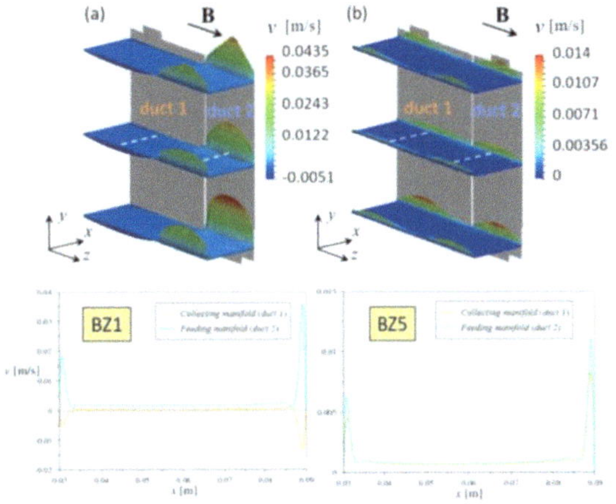

Figure 6 Results for Ha=4000 and u_0=0.1mm/s, at the bottom of the module in BZ1 (a), in the middle in BZ5 (b). On the upper half of the figure, 3D views of velocity distribution are displayed at three poloidal locations. On the lower half, profiles of vertical velocity are plotted along the radial x coordinate, as indicated by the white dashed lines.

In Figure 7a the pressure is plotted as a function of the vertical poloidal direction y along the central line of the ducts forming the manifolds for the flow at Ha=4000 and u_0=0.1 mm/s. The distance between the two curves, marked by red arrows, indicates the pressure difference between feeding and draining manifold, i.e. the available pressure heads between entrance and exit of the corresponding breeder zones. This pressure difference is required to overcome the pressure drop across the windows in the back plate (BP) that connect manifold and BUs, and the one along the breeder zone. Experiments performed to investigate MHD flows in a scaled mock-up of a HCLL blanket showed that the pressure drop in the BUs, is almost negligible compared to the other contributions in the blanket module [8]. The additional pressure drops present at the connections between two portions of the manifolds, i.e. the steps visible in the pressure distribution in Figure 7a, result from the local 3D MHD effects that occur when the liquid metal expands/contracts along magnetic field lines. For the flow at Ha=4000 and u_0 =0.1 mm/s the total pressure drop in the feeding (FM) and collecting (CM) manifolds is given by Δp_{FM}=0.092 bar, Δp_{CM}=0.048 bar, respectively. The pressure drop in the draining manifold is smaller than the one in the feeding channel because its cross-section is larger.

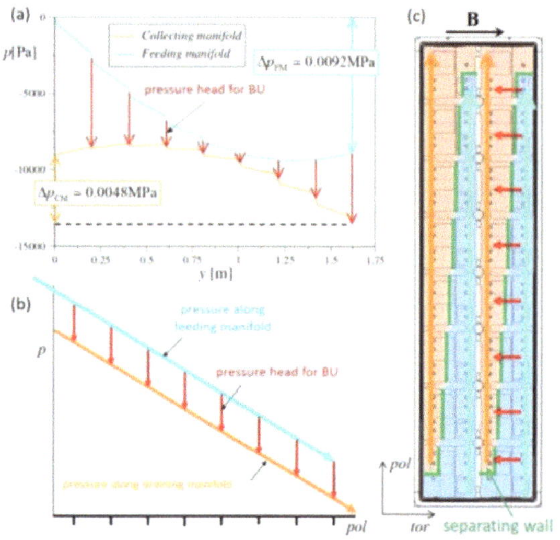

Figure 7 (a) Pressure plotted along the vertical poloidal direction in feeding and draining manifolds for Ha=4000 and u_0=0.1 mm/s. (b) Schematic ideal linear profile of poloidal distribution of pressure in the manifold. (c) Sketch of PbLi path and proposed modification of the manifold design.

In an ideal situation, the pressure heads for each breeding zone should be equal in order to guarantee a uniform flow partitioning among all BUs. However, this would be possible only if the pressure along both manifolds varied linearly and with same slope, as sketched in Figure 7b. In the present manifold design, the cross-sections of the two ducts differ from each other and remain constant along the poloidal length of the module. Consequently, the mean velocity in the feeding manifold becomes smaller along the poloidal direction, when the liquid metal flows from BZ1 to BZ8. Therefore, the pressure gradient decreases. Instead, in the collecting manifold the flow rate increases along the liquid metal flow path as well as the pressure gradient. The larger pressure head at the external BUs suggests that the flow rate in central BUs is considerably reduced compared to the one at the ends of the module.

Such behavior has been studied in the past by experiments performed in a mock-up of a HCLL TBM [9] and design modifications had been proposed to overcome this undesirable effect. A more uniform partitioning of the liquid metal among the breeding zones could be achieved by adapting the manifold cross-sections to the changing flow rates, so that a constant average velocity is maintained along the liquid metal path, and same pressure head could be reached for each BU. In order to realize these conditions, the vertical wall separating the two manifolds should be moved stepwise in toroidal direction, as proposed in Figure 7c, where the new wall position is drawn in green.

The present numerical study allows for a first estimation of pressure drop in the feeding and in the draining manifolds of a WCLL TBM. Numerical results confirm the typical scaling for the pressure drop in inertialess MHD flows under strong magnetic fields, according to which pressure losses scale linearly with the mean velocity and quadratically with the magnetic field strength, $\Delta p \sim \sigma L B^2 u_0$. However, it is still necessary to consider that, even if the pressure drop in the manifolds represents a large part of the total pressure drop in the PbLi system, other contributions should be estimated as well. For instance pressure drops in the long piping system supplying PbLi to the TBM were not part of the present investigation. In these pipes PbLi flows under the action of a spatially varying magnetic field and the velocity is much higher than in the BUs.

The obtained results show that the pressure drop in the manifolds is acceptable for ITER applications. However, a more important aspect is the non-uniform partitioning of the liquid metal flow among different breeding zones due to the geometry of the manifolds in which the cross-sections do not adjust to the changing flow rates. The proposed modification of the manifold design will be evaluated in future studies.

Design and fabrication of a WCLL TBM mock-up for MHD experiments

For MHD experiments in the MEKKA laboratory a mock-up of the ITER WCLL TBM has been designed and fabricated (see Figure 8). Objective of these experiments is to achieve knowledge about MHD pressure drop, when the liquid metal flows around obstacles (water pipes) and through contractions and expansions created by stiffening plates in the manifolds and to determine the distribution of PbLi flow among breeder units.

The WCLL mock-up has been 1:2.5 downscaled to fit into the gap of the magnet at the MEKKA laboratory. Figure 9 shows a transparent CAD view of the mock-up for visualization of the PbLi flow paths.

Figure 8 Manufactured WCLL TBM mock-up.

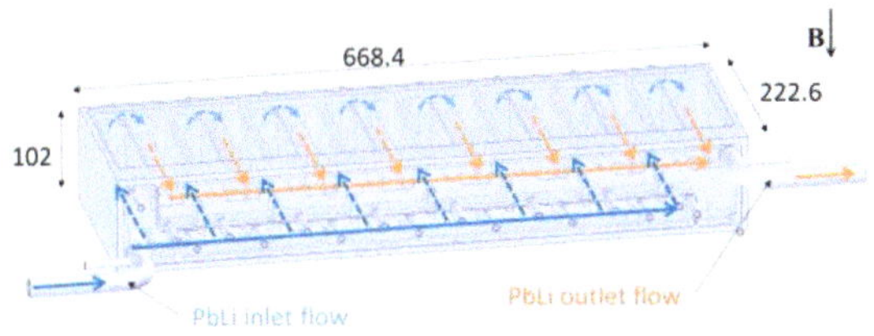

Figure 9 Transparent CAD view with main dimensions (in mm) and PbLi flow paths.

The WCLL mock-up has been simplified to meet manufacturing concerns and to reduce fabrication costs, while keeping the geometry for the liquid metal flow as foreseen in [10]. An exploded CAD view in Figure 10 shows the major parts of the mock-up. The main body consists of a single piece, which forms the walls of all breeder units (BUs), including first wall, stiffening plates, back plate and walls separating the two inlet and outlet manifolds for the PbLi flow. Manifolds are closed by two back plates. Details concerning the water channels inside the blanket walls have been omitted since it should not affect the liquid metal flow. Cooling pipes are all present in the mock-up and simulated by solid dummy elements. All parts are made of 1.4571 austenitic steel that has good compatibility with the used model fluid NaK.

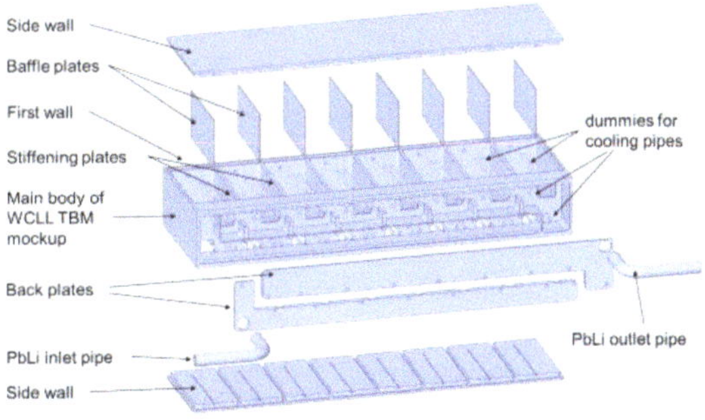

Figure 10 Exploded view of the WCLL MHD mock-up with major components: main body, side walls, back plates as cover for the manifolds, and dummy parts that mimic the water pipes.

The liquid metal is fed into the manifold of the main body and removed from it through circular pipes, which are connected to the existing loop in the laboratory. The PbLi is distributed and collected by two manifolds into and out of the BUs. Each BU is fed through a small window in the back wall of the manifolds. The PbLi flow is redirected at the first wall and guided back into the outlet manifold (see Figure 9).

Pressure differences will be measured between several points of the mock-up using a piping system between pressure taps and the pressure transducers which are located outside of the magnet. Positions of pressure taps have been chosen to also satisfy the function of feeding, draining and venting (see Figure 12).

Flows in different parts of the module are electrically coupled since currents induced in one part may leak across electrically conducting walls into neighboring channels. Electromagnetic flow coupling and flow partitioning among BUs is investigated by the electric potential distribution on the walls. Experiments using this mock-up in the MEKKA laboratory will start in the second half of 2021.

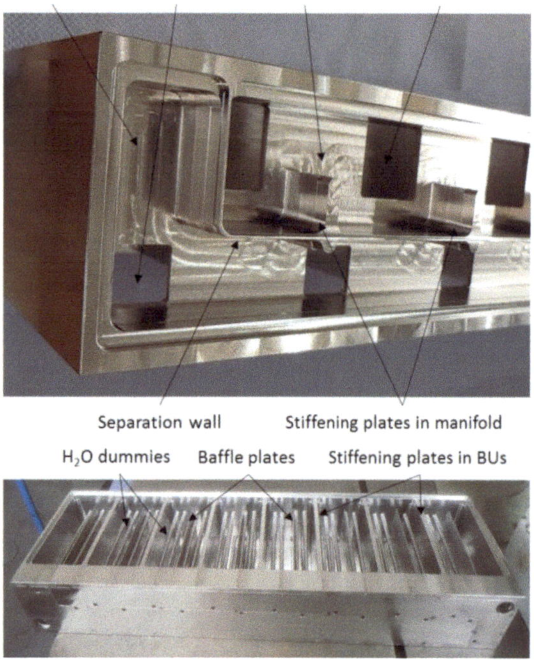

Figure 11 WCLL TBM mock-up: inlet and outlet manifolds (top), breeder zone with dummies representing cooling pipes, baffle plates and stiffening plates (bottom).

Figure 12 Finalized mock-up.

Reconstruction of 3D velocity profiles from measured surface potential data

Liquid metal pipe flow entering a strong magnetic field had been investigated experimentally and theoretically in a previous campaign [11] [12] using the MEKKA facility [13]. Experimental data for electric potential Φ on the external surface of the pipe had been recorded and published in [11]. This data displayed in Figure 13 has been analyzed and potential values ϕ on the fluid-wall interface have been evaluated by solving a sort of inverse problem across the thick-walled pipe. Potential at the fluid-wall interface is of particular interest since it is a result of the interaction of the flow with the magnetic field and from its knowledge it is possible to derive an approximate reconstruction of the velocity distribution in the entire fluid volume.

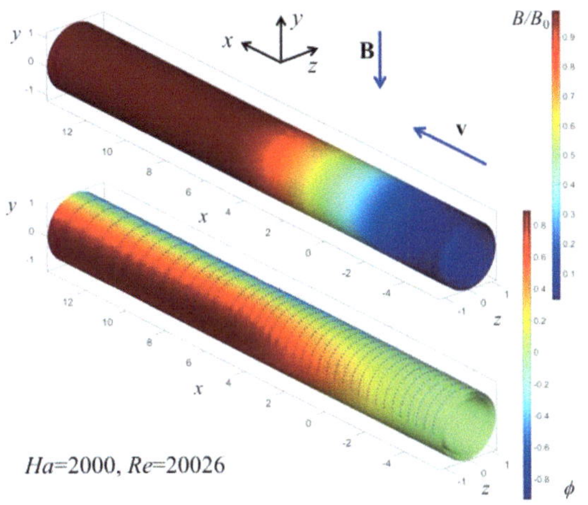

Figure 13 Distribution of magnetic field $B(x)$ projected to the pipe surface and contours of nondimensional electric potential Φ measured on the external surface of the pipe. The positions of the potential sensors are marked by blue dots.

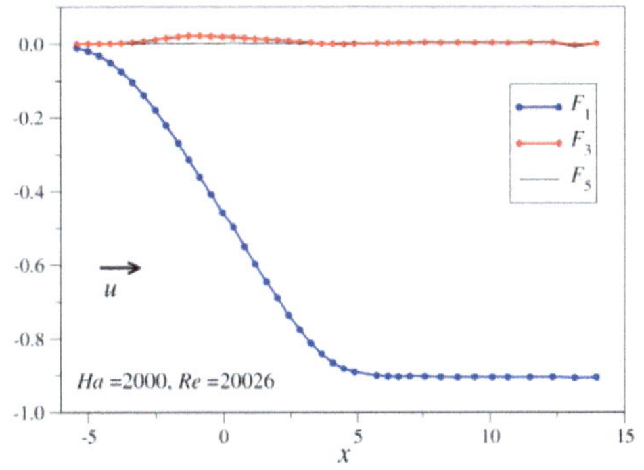

Figure 14 Fourier coefficients $F_k(x)$ of surface potential $\Phi(\alpha,x)$ at the external surface of the pipe for Ha=2000, Re=20000.

A first step in the analysis is a decomposition of surface data in circumferential Fourier modes

$$\phi(R,\alpha,x) = \Phi(\alpha,x) = \sum F_k(x)\sin(k\alpha),$$

where the nondimensional radius R denotes the ratio of external to internal radius of the pipe and α stands for the angle in a polar coordinate system, measured with respect to the magnetic field direction. The F_k are obtained as best fit of the approach introduced above with the experimental data. Results of $F_k(x)$ displayed in Figure 14 show that the major physical information is already contained in the first 3 modes while values of mode 5 are already negligibly small.

Assuming a balance between flow-induced electric field $\mathbf{v}\times\mathbf{B}$ and potential gradient $\nabla\phi$, and the fact that the fluid core potential is constant along magnetic field lines [14], we can express the axial component of velocity as

$$u = B^{-1}\frac{\partial\phi}{\partial z} = \frac{1}{B\cos\alpha}\frac{\partial\phi}{\partial\alpha},$$

when ϕ is known in terms of its Fourier expansion. Figure 15 shows velocity contours for the two Hartmann numbers Ha=2000 and Ha=5485. We observe the formation of increased velocities near the sides $z\approx\pm1$ and reduced velocity in the center, in the region where the magnetic field starts rising.

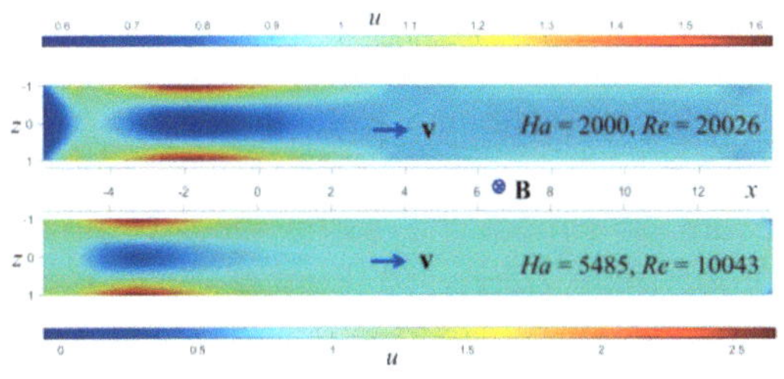

Figure 15 Contours of axial velocity $u(x,z)$ estimated from measured wall potential for Ha=2000, Re=20026 and Ha=5485, Re=10043.

The distributions of axial velocity at the sides $u(x,z=\pm1)$ and in the center $u(x,z=0)$ are plotted in Figure 16. For $x<-5$, both the magnetic field and induced potentials are pretty small so that data evaluation for velocity becomes difficult and the derived data might become unreliable (larger scatter of data for $x<-5$). However, for $x>-5$ the data shows clear trends. Maximum values reach up to $u_{max}=1.62$ and 2.64 at $x\approx-1.68$ and $x\approx-3.31$ for $Ha=2000$ and $Ha=5485$, respectively.

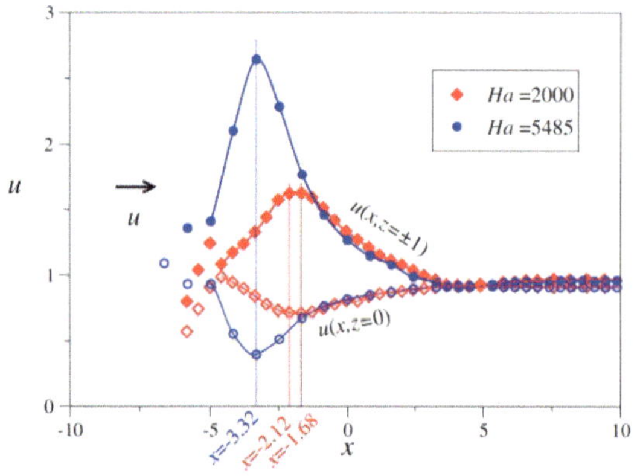

Figure 16 Velocities near the sides (filled symbols) $u(x,z=\pm1)$ and in the center (open symbols) $u(x,z=0)$ for two Hartmann numbers.

It can be observed that the axial position of maximum jet velocity near the sides shifts upstream as the magnetic field B_0 increases. This is most probably caused by the fact that locally the magnetic field increases earlier for higher Hartmann numbers. In order to support this assumption we compare the two experiments in terms of their local Hartmann numbers $Ha(x)=Ha\,B(x)/B_0$ as shown in Figure 17, where the two positions of maximum jet velocity have been added to the figure. It can be seen that in both cases the position of maximum jet velocity occurs at almost the same local Hartmann number near $Ha(x)\approx560$. Further downstream, the velocities near the sides decrease slowly and approach that of a uniform core flow with $u\approx0.92$.

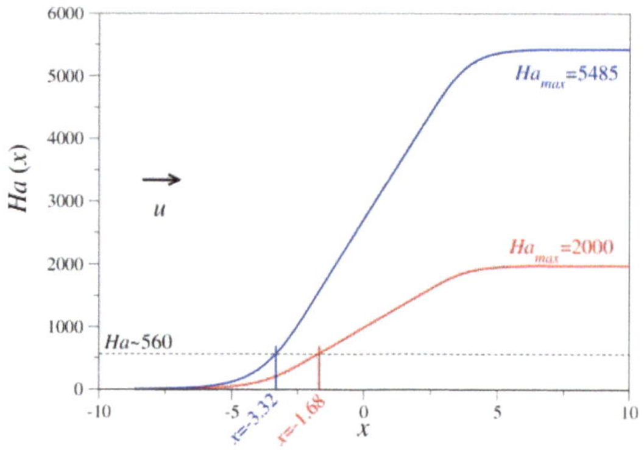

Figure 17 Distribution of local Hartmann number $Ha(x)$ for the two cases considered. The marked positions $x=-3.32$ and $x=-1.68$ denote locations of maximum jet velocity.

The transverse distribution of velocity is shown in Figure 18 for the high Hartmann number $Ha=5485$ and $Re=10043$. We observe high velocity magnitudes near the sides at axial positions where the magnetic field starts rising with a reduced velocity in the center. Further downstream the flow gradually becomes uniform as the magnetic field approached its uniform high magnitude.

The advantage of the presented method for evaluation of velocity from measured potential data is that it yields a global overview of velocity in the entire flow field in contrast to invasive measurements where a probe is traversed along a single line of a cross-section. More details and a discussion about the accuracy of the procedure can be found in [15].

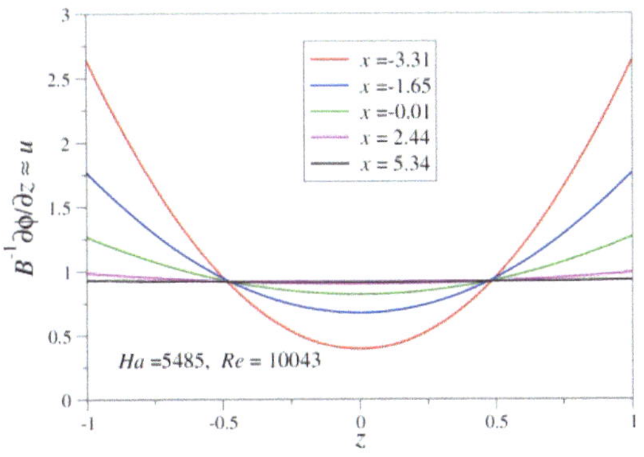

Figure 18 Transverse profiles of axial velocity at different axial positions for Ha=5485 and Re=10043.

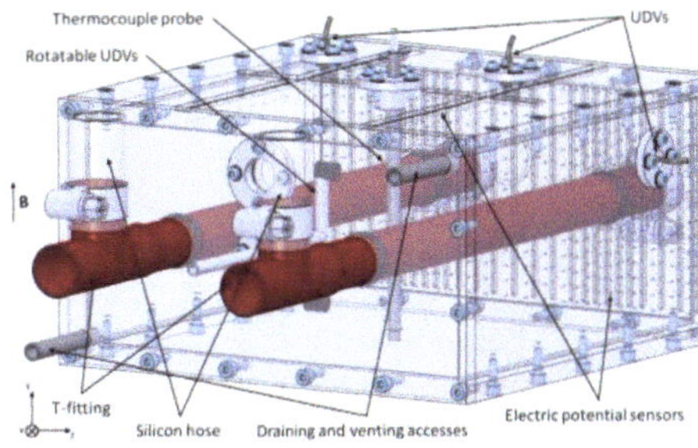

Figure 19 Sketch of the test section.

Experimental study of heat transfer in magneto-convective flows related to WCLL blankets

In WCLL blankets, water-cooled pipes are immersed in the liquid metal to extract the heat from the breeding zone. These pipes generate large thermal gradients within the liquid metal that may lead to quite complex flows governed by the combined effect of buoyancy, electromagnetic force, and pressure gradient. Although some characteristic features associated with such flow have already been identified by numerical studies [16] [17] experimental investigations of magneto-convective flow in prototypical geometries are lacking. Yet, additional studies are required to improve the understanding of this type of flow and support the complex design of a WCLL blanket. To this end, a simplified model geometry was defined with the aim of providing a first experimental database of MHD buoyant flows with heat transfer in WCLL-like geometries.

A sketch of the test section designed to perform magneto-convection experiments in the MEKKA laboratory is shown in Figure 19. It consists of the 200×200×100 mm rectangular cavity made of PEEK plastic through which two parallel horizontal pipes are inserted. Both pipes are connected to their own temperature-controlled water circuits such that they can be kept at constant temperatures T_1 and T_2 to generate the horizontal temperature gradient driving the flow. The pipes are made of copper and they have an inner core to ensure as good isothermal conditions as possible. They are coated with a thin layer (2 µm) of silicon carbide that provides electrical insulation and prevents corrosion that could occur at higher temperatures in contact with eutectic gallium-indium-tin, which is used as model fluid. The entire test section is instrumented with 71 thermocouples, 416 electrodes and 3 UDV transducers to record simultaneously temperatures, electric potentials at the walls, and velocities. Additionally, flowrates and temperature drop/rise along both pipes are also monitored to evaluate the amount of heat exchanged between the pipes. More details about the test section and the instrumentation can be found in [18].

Figure 20 Photo of the instrumented test section partially thermally insulated before insertion in the magnet gap of the MEKKA laboratory.

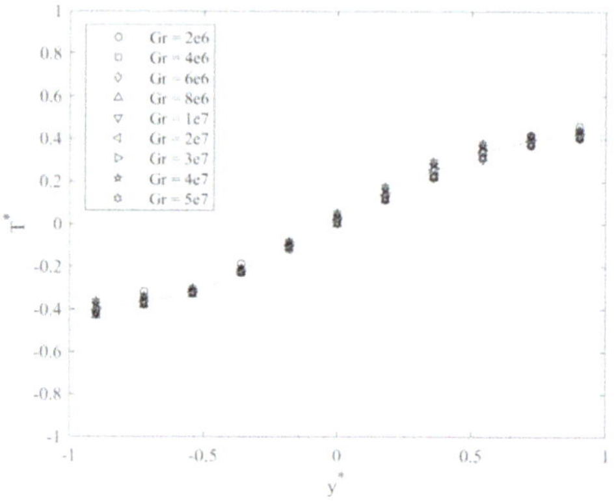

Figure 21 Nondimensional temperature profiles measured at the center of the test section ($x^* = -0.13$ & $z^* = 0$) for $Ha = 0$.

Experiments were performed for Hartmann numbers up to 3000 and Grashof numbers ranging from $2.0 \cdot 10^6$ to $1.0 \cdot 10^8$, corresponding to temperature differences $\Delta T = T_2 - T_1$ imposed between the pipes varying between 1.4 °C and 69.5°C. Whenever possible, the temperature of the pipes was chosen such that average temperature of the liquid metal $\bar{T} = (T_2 + T_1)/2$ was close to the temperature inside the magnet gap – set to 30 °C – in an effort to limit as much as possible the external heat exchanges.

The temperature distributions were measured at the center of the box, on the middle of the sidewall (on the heated side, $z^* = -2.00$) and on the back plate ($x^* = -1.98$), as well as on the middle of the top cover ($y^* = 1.00$) and in both pipes. Results obtained are scaled and plot as nondimensional temperature T^* defined as follow:

$$T^* = \frac{T - \bar{T}}{\Delta T/2}$$

For hydrodynamic experiments, performed in the absence of magnetic field ($Ha = 0$), a convection cell forms in the center of the cavity, between the two pipes. The buoyant flow results in a thermal stratification with the hot fluid staying on the top and the cold fluid on the bottom as seen on the non-dimensional temperature profile measured at the center probe shown in Figure 21.

Magnetohydrodynamic experiments exhibit quite different temperature distributions depending on the strength of the applied magnetic field. At the lowest Ha numbers, the turbulence suppression by the magnetic field first results in larger temperature gradient within the flow (Figure 22), i.e. the thermal stratification becomes stronger. As the magnetic field further increases, the convection motion is damped by the Lorentz forces induced in the liquid metal. Consequently, the vertical thermal gradient at the center of the cavity progressively diminishes until the horizontal thermal stratification has completely disappeared. Eventually, convection motion is so damped that thermal conduction emerges as the dominating heat transfer mechanism and the temperature distribution becomes symmetric with respect to the vertical middle plane of the test section at the highest Ha numbers.

This result is confirmed by the measurement of the heat transfer coefficient. The intensity of the convective heat transfer has been quantified by deriving the amount of heat released/absorbed at the hot/cold walls, respectively, from the energy balance in both water pipes. The heat fluxes at the pipes can be then evaluated knowing the exchange area and expressed in nondimensional form as Nusselt number

$$Nu = \frac{\dot{m} c_p \Delta T_x L}{\pi d l k \left(T_{1,2} - \bar{T}\right)}$$

Here, $\dot{m}$ denotes the mass flow rate of water in the pipe and ΔT_x the temperature drop/rise. k and c_p are the thermal conductivity of galinstan and the specific heat of water, respectively, and d and l are the diameter and length of the pipes (see Figure 23). The characteristic length of the problem L considered here is half distance between the pipes which is equal to the half size of the box measured along magnetic field lines (Hartmann length).

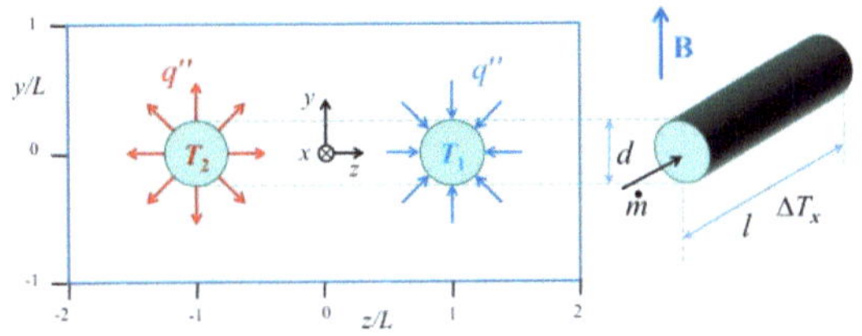

Figure 23 Sketch for the definition of parameters used to derive the Nusselt number.

The results are presented in Figure 24 where Nu is plotted against the quantity Gr/Ha^2 representing the ratio of the buoyancy force over the electromagnetic force. The competition between buoyancy forces, promoting the convection, and the electromagnetic forces, suppressing the flow, lead to 2 distinct heat transfer regimes:

- For $Gr/Ha^2 < 10$, Nu is constant and close to unity indicating that heat transfer is purely conductive. Almost all convection motion has been suppressed by the Lorentz force.
- For $50 < Gr/Ha^2 < 4000$, the intensity of heat transfer increases with Gr/Ha^2. In this regime, a scaling law has been established such as $Nu \propto \left(Gr / Ha^2\right)^{1/3}$.

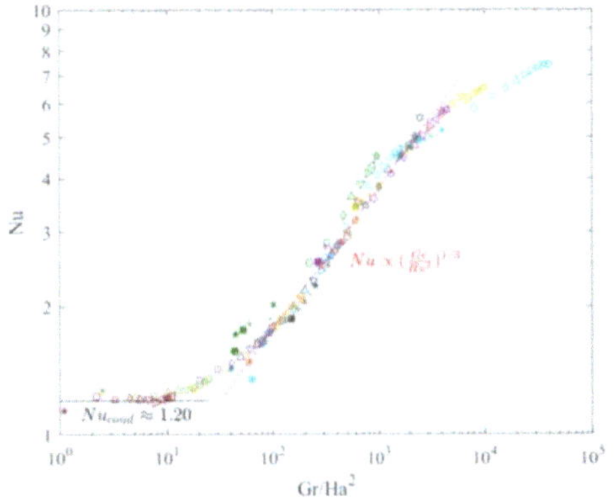

Figure 24 Nusselt numbers measured for several Grashof and Hartmann numbers.

As Gr/Ha^2 grows further, the pace at which Nu increases slows down and eventually the flow becomes turbulent as proven by current experimental results. While this work already provides valuable data, extended investigation will be carried out focusing on the onset of turbulence in magneto-convective flow.

Further work

In addition to the topics described above, the MHD group at ITES KIT contributed in the reporting period 2020 with scientific papers to the

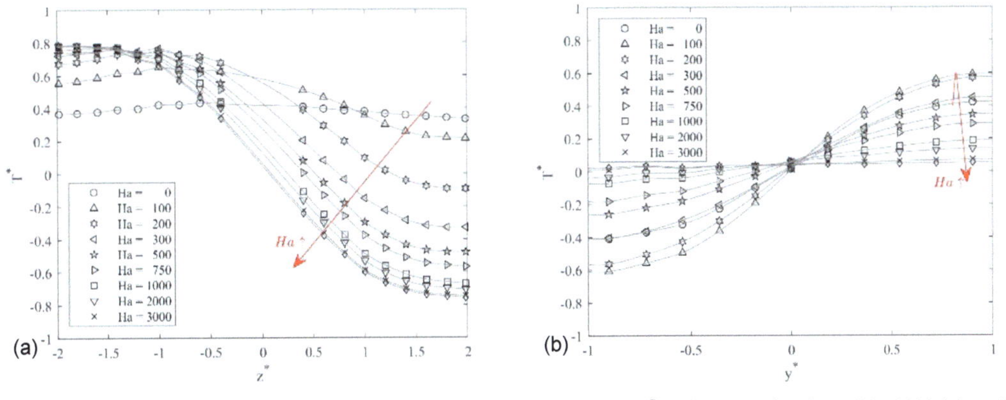

Figure 22 Nondimensional temperature profiles measured for $Gr = 2 \cdot 10^7$ and Ha ranging from 0 to 3000 (a) on the middle of the top wall ($x^* = -0.13$ & $y^* = 1$) and (b) at the center of the test section ($x^* = -0.13$ & $z^* = 0$).

development of fusion technology [19], [20], [21], [22], [23], [24], [25].

Acknowledgement: This work has been carried out within the framework of the EUROfusion Consortium and has received funding from the Euratom research and training programme 2014-2018 and 2019-2020 under grant agreement No 633053. The views and opinions expressed herein do not necessarily reflect those of the European Commission.

References

[1] Malang, S.; Einrichtung zur Verringerung des MHD-Druckverlustes in dickwandigen Kanälen, Patent DE 36 00 645 C2, 1987.

[2] Miyazaki, K.; Nakano, M.; Horiba, T.; Inoue S. and Yamaoka, N.; "MHD pressure drop along a NaK flow around a stepwise change of tube diameter under a transverse magnetic field," *Fusion Engineering and Design,* vol. 8, pp. 233-239, 1989.

[3] Bühler, L.; Mistrangelo C. and Brinkmann, H.-J.; "Experimental investigation of liquid metal MHD flow entering a flow channel insert," *Fusion Engineering and Design,* vol. 154, p. 111484, 2020.

[4] Federici, G.; Boccaccini, L.; Cismondi, F.; Gasparotto, M.; Poitevin Y. and Ricapito, I.; "An overview of the EU breeding blanket design strategy as an integral part of the DEMO design effort," *Fusion Engineering and Design,* vol. 141, pp. 30-42, 2019.

[5] Bühler, L.; "A parametric study of 3D MHD flows in expansions of rectangular ducts," Fusion *Science and Technology,* vol. 52, no. 3, pp. 595-602, 2007.

[6] Mistrangelo, C. and Bühler, L.; "Development of a numerical tool to simulate magnetohydrodynamic interactions of liquid metals with strong applied magnetic fields," Fusion Science and Technology, vol. 60, no. 2, pp. 798-803, 2011.

[7] Smolentsev, S.; Badia, S.; Bhattacharyay, R.; Bühler, L.; Chen, L.; Huang, Q.; Jin, H.-G.; Krasnov, D.; Lee, D.-W.; Mas de les Valls, E.; Mistrangelo, C.; Munipalli, R.; Ni, M.-J.; Pashkevich, D.; Patel, A.; Pulugundla, G.; Satyamurthy, P.; Snegirev, A.; Sviridov, V.; Swain, P.; Zhou T. and Zikanov, O.; "An approach to verification and validation of MHD codes for fusion applications," *Fusion Engineering and Design,* vol. 100, pp. 65-72, 2015.

[8] Mistrangelo, C. and Bühler, L.; "Magnetohydrodynamic pressure drops in geometric elements forming a HCLL blanket mock-up," *Fusion Engineering and Design,* vol. 86, pp. 2304-2307, 2011.

[9] Bühler, L.; Mistrangelo, C.; Brinkmann, H.-J. and Koehly, C.; "Pressure distribution in MHD flows in an experimental test-section for a HCLL blanket," *Fusion Engineering and Design,* vol. 127, pp. 168-172, 2018.

[10] Batal, T.; *Assembly of WCLL TBM: CAD product and parts (by CEA).,* personal communication November 27, 2019.

[11] Bühler, L.; Brinkmann, H.-J. and Mistrangelo, C.; "Experimental investigation of liquid metal pipe flow in a strong non-uniform magnetic field," *Magnetohydrodynamics,* vol. 56, pp. 81-88, 2020.

[12] Klüber, V.; Mistrangelo, C. and Bühler, L.; "Numerical simulation of 3D magnetohydrodynamic liquid metal flow in a spatially varying solenoidal magnetic field," *Fusion Engineering and Design,* vol. 156, p. 111659, 2020.

[13] Barleon, L.; Mack, K.-J. and Stieglitz, R.; "The MEKKA-facility a flexible tool to investigate MHD-flow phenomena," 1996.

[14] Picologlou, B. F. and Reed, C. B.; "Experimental investigation of 3-D MHD flows

at high Hartmann number and interaction parameter," in *Liquid Metal Magnetohydrodynamics*, J. Lielpetris and R. Moreau, Eds., Dordrecht, Kluwer, 1989, pp. 71-77.

[15] Bühler, L. ; Lyu, B.; Brinkmann, H.-J. and Mistrangelo, C.; "Reconstruction of 3D MHD liquid metal velocity from measurements of electric potential on the external surface of a thick-walled pipe," *Fusion Endineering and Design,* vol. 168, p. 112590, 2021.

[16] Bühler, L. and Mistrangelo, C.; "MHD flow and heat transfer in model geometries for WCLL blankets," *Fusion Engineering and Design,* vol. 122, pp. 919-923, 2017.

[17] Tassone, A.; Caruso, G.; Giannetti, F. and Nevo, A. D.; "MHD mixed convection flow in the WCLL: heat transfer analysis and cooling system optimization," *Fusion Engineering and Design,* vol. 146, 2019.

[18] Koehly, C.; Bühler, L. and Mistrangelo, C.; "Design of a test section to analyze magneto-convection effects in WCLL blankets," *Fusion Science and Technology,* vol. 75, pp. 1010-1015, 2019.

[19] Klüber, V.; Bühler, L. and Mistrangelo, C.; "Numerical investigation of liquid metal flow in square channels under inclined magnetic fields for fusion relevant parameters," *Magnetohydrodynamics,* vol. 56, pp. 149-156, 2020.

[20] Mistrangelo, C.; Bühler, L.; Smolentsev, S.; Klüber, V.; Maione, I. and Aubert, J.; "MHD flow in liquid metal blankets: major design issues, numerical analysis and code validation," in *Invited talk at the 31st Symposium on Fusion Technology (SOFT2020)*, Virtual edition, 2020.

[21] Mistrangelo, C.; Bühler, L. and Klüber, V.; "Three-dimensional magneto convective flows in geometries relevant for DCLL blankets," *Fusion Engineering and Design,* vol. 159, p. 111686, 2020.

[22] Mistrangelo, C.; Bühler, L.; Smolentsev, S.; Klüber, V.; Maione, I. and Aubert, J.; "MHD flow in liquid metal blankets: major design issues, MHD guidelines and numerical analysis," *Fusion Engineering and Design,* p. accepted, 2021.

[23] Mistrangelo, C;. Bühler, L.; Koehly, C. and Ricapito, I.; "Magnetohydrodynamic velocity and pressure drop in WCLL TBM," in *Proceedings of the 28th IAEA Fusion Energy Conference, May 10-15, Virtual*, 2021.

[24] Mistrangelo, C.; Bühler, L.; Koehly, C. and Ricapito, I.; "Magnetohydrodynamic velocity and pressure drop in WCLL TBM," *Nuclear Fusion,* p. in print, 2021.

[25] Smolentsev, S.; Rhodes, T.; Yan, Y.; Tassone, A.; Mistrangelo, C.; Bühler, L. and Urgorri, F. R.; "Code-to-code comparison for a PbLi mixed-convection MHD flow," *Fusion Science and Technology,* vol. 76, pp. 653-669, 2020.

Group: Severe Accident Research

SAR/ITES research activities in 2020

Xioayang Gaus-Liu

In 2020, the research activities of ITES/SAR were focused on the analyses and mitigation of severe accident of boiling water reactor (BWR) and the construction of a new facility ESFR-LIVE as a task in the EU Horizon2020 project ESFR-SMART.

In-vessel melt and debris retention is also a defense-in-depth strategy during the course of a severe accident scenario of BWR. The Fukushima-Accident reminds the urgent necessity of a melt retention concept specialized for the BWR reactor design. External cooling by flooding the reactor cavity, the most studied one in the past, could be functional for Pressurized Water Reactors (PWR), characterized by a small reactor pit and low position of the Reactor Pressure Vessel (RPV) in the cavity. It is not suitable for BWR with a deep und large cavity. On the other hand, the lower plenum of a BWR has a dense array of control rod guide tubes (CRGTs), which penetrate the RPV bottom vertically to the cavity. CRGTs can contribute to the cooling of a relocated debris bed or a molten pool in the lower plenum in two ways: a) cooling inside the corium by the water-cooled CRGTs and b) directing water via CRGTs to the top surface of the corium to enable top cooling. The cooling effects of CRGTs both in a non-molten debris bed and in a molten pool were experimentally simulated in the LIVE3D test facility [1]. The test configurations simulated different scenarios: no cooling at all, cooling only by CRGTs and cooling by CRGTs together with external wall cooling. 16 CRGTs are simulated and were arranged in 4 groups. The tubes were mounted on an upper lid with water distribution system, and inserted vertically approaching the bottom of the vessel. Figure 1 shows the tube inside.the vessel and the water distribution system at the upper lid.

Figure 1. LIVE-BWR tests with simulants of CRGTs.

The heating and cooling of debris were performed in 6 phases, in each of which there was a high power period and a low power period, as shown by Ptotal in Figure 4, right diagram. The external water cooling were only performed in the phase 5 and phase 6 with the combination of CRGT cooling. During the high power period, the CRGT cooling could only remove a part of the power, with the consequence that the debris temperature arise continuously. During the low power period, the CRGT tubes could effectively reduce the debris bed temperature and mitigate the thermal load on the vessel wall. The cooling of CRGT tubes was more effective in a molten pool, as shown in the Figure 2, bottom. The heat removal rate via CRGT is as effective as the external cooling.

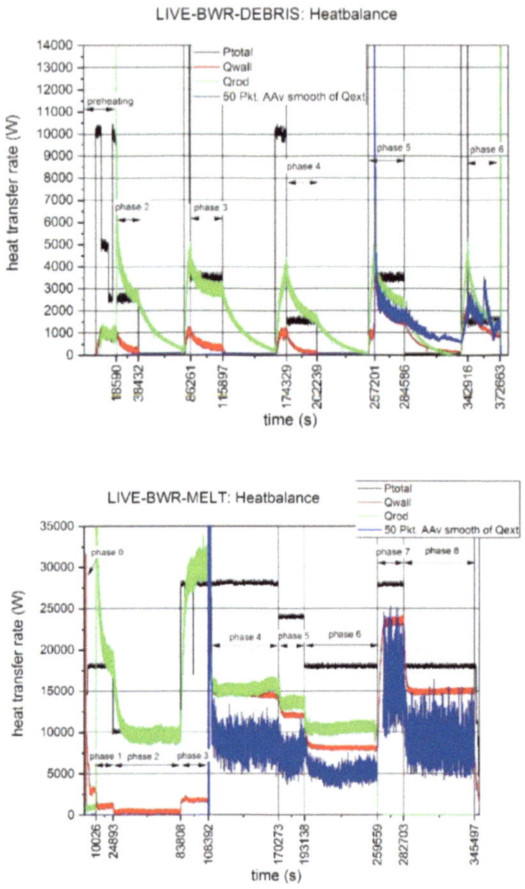

Figure 2. Power input (Ptotal) and heat transfer via CRGT tubes (Qrod) and via external cooling (Qwall). Top: Debris test, Bottom: molten pool test.

Another experiment was the investigation in melt-down of a debris bed with heterogeneous materials. This experiment was assigned by the JAEA /CLAD laboratory with the aim to understand the accident scenario of Unit 2 Fukushima accident and thus accomplish safe decommissioning of the Fukushima Daiichi Nuclear Power Station [2]. Till present, few experiments investigate a meltdown process of debris bed consisting of materials with different melting temperatures. Two experiments were performed on the LIVE3D facility: LIVE-J1 and LIVE-J2 tests. LVE J1 investigates the melt down of debris, whereas LIVE-J2 investigates a corium of debris and liquid melt with and without external cooling.

The important outcome of LIVE-J1 tests are a) the ongoing shrinkage of debris bed volume and height during the melt-down of the particles with low melting point; b) the shift of hotspot in the debris bed and thermal load on the vessel wall from bottom toward the upper region during the melting. Figure 3 shows the debris bed before and after meltdown process.

Figure 3. Appearance of initial debris bed (top) and the debris bed after the test (bottom) in LIVE-J1 test.

The third accomplishment of SAR group in 2020 is the design and construction of a new large-scale 3D vessel, ESFR-LIVE imitating the geometry of the core-catcher of a sodium cooled fast reactor (SFR) in a diameter scale of 1:6 [3]. The core catcher has the function to collect the relocated melt derived from the upper core area and to prevent local thermal attack of melt on the reactor vessel. The thermodynamics of the melt with decay heat and its heat transfer at the core catcher boundaries will be studied experimentally at KIT and numerically at CEA. The facility will be in commissioning in middle of 2021. The ESFR-LIVE test vessel simulates the typical tray geometry of an in-vessel SFR core catcher, whose cavity has a geometry of truncated cone in the lower part, and cylinder in the upper part. The inner diameter of the upper cylinder is 1 m.

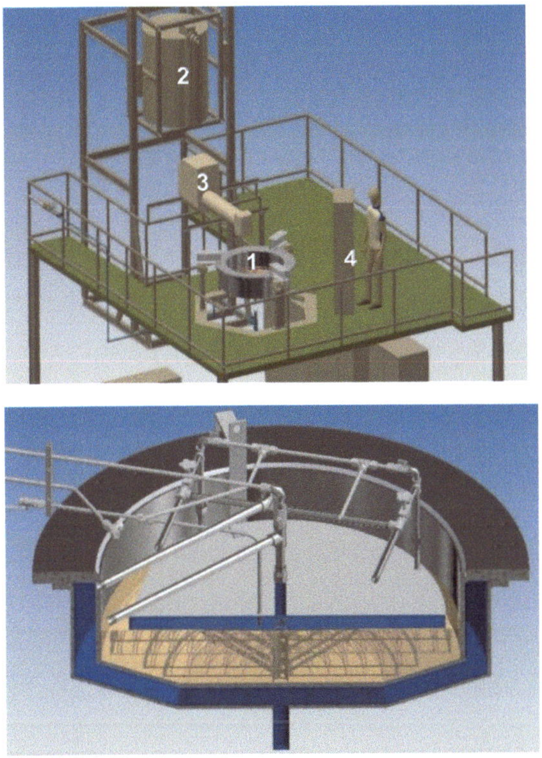

Fig.4. Top: LIVE experimental platform with ESFR-LIVE facility. 1- ESFR test vessel, 2 - heating furnace for melt preparation, 3 - pouring spout, 4 - instrumentation board. Bottom: ESFR-LIVE test vessel with all boundary cooling system.

Four planes of resistance heating element are designed to simulate the decay heat of the corium. Two planes are located in the truncated cone (HE1 and HE2) and the other two are located in the cylindrical part (HE3 and HE4). The upper two heating planes can be removed individually to enable different pool heights at the top boundary. The total heating power can reach 86.5 kW. The flexibility of the heating planes enable a large variety of the pool geometries, e.g. possible by shutdown the one or two of the lower heaters. Correspondingly, bottom crust grows upwards till the level of the lowest heater in operation. The high power provided also make it possibly for experiments with high decay power, such as debris dryout experiments.

Two experimental series in ESFR-SMART project are foreseen in 2021. The experimental data will be used for the validation of the numerical calculations with TriCAD code in CEA and for analytical analysis done by KIT and CEA.

To be published:

[1] Fluhrer, B.; Gaus-Liu, X.; Cron, T.; Stängle, R.; Vervoortz, M.; Wenz, T.; LIVE-BWR EXPERIMENTS TO STUDY THE EFFECT OF CRGT COOLING ON IN-VESSEL MELT RETENTION STRATEGY, The 19th International Topical Meeting on Nuclear Reactor Thermal Hydraulics (NURETH-19), Brussels, Belgium, March 6 - 11, 2022

[2] Madokoro, H.; Gaus-Liu, X.; et.al.; LIVE-J1 EXPERIMENT ON DEBRIS MELTING BEHAVIOR TOWARD UNDERSTANDING LATE IN-VESSEL ACCIDENT PROGRESSION OF THE FUKUSHIMA DAIICHI NUCLEAR POWER STATION, The 19th International Topical Meeting on Nuclear Reactor Thermal Hydraulics (NURETH-19), Brussels, Belgium, March 6 - 11, 2022

[3] Gaus-Liu, X. ; Bigot, B.; Journeau, C.; Payot, F.; Cron, T.; Clavier, R.; Peybernes, M.; Angeli, P. E.; Fluhrer, B.; Experiment and Numerical Simulations on SFR Core-catcher Safety Analysis after Relocation of Corium, CN291-IC336, #FR22, International Conference on Fast Reactors and Related Fuel Cycles Fr22: Sustainable Clean Energy for the Future (CN-291), 25-28 April 2022, Beijing, China.

Group: Multiphase Flow

Detailed investigations on flow boiling of water up to the critical heat flux

Stephan Gabriel, Giancarlo Albrecht, Wilson Heiler, Felix Heineken

Introduction

The prediction of two-phase flows with heat transfer using system codes and CFD methods is essential in various power-engineering applications. The current projects of the working group Multiphase Flow (MPF) deal with topics from the field of In Vessel Retention (IVR) by External Reactor Vessel Cooling (ERVC) as an emergency procedure as well as safety-related evaluation of different designs of small modular reactors (SMR) with respect to the boiling processes under full-scale conditions.

The research group concentrates on experimental investigations, in particular on boiling under forced convection up to the critical heat flux. The activities in 2020 focused on the measurement of the critical heat flux at fluctuating mass flows, which can occur under ERVC conditions at the low-pressure COSMOS-L facility. At COSMOS-H , work was continued on completion and commissioning of the loops and, in addition, an initial project relating to small modular reactors (SMR) was started. Within the scope of both projects, the measurement technology was further developed in order to be able to record the relevant measured variables in as much as detailed as possible.

COSMOS-L

Within the scope of the BMWi - project KEKSIMA the critical heat flow under periodically fluctuating mass flows is being investigated in comparison to steady state conditions. The aim is a better understanding of the physical processes in IVR/EVRC scenarios as they are for example conceivable in Gen III+ reactors (e.g. AP1000).

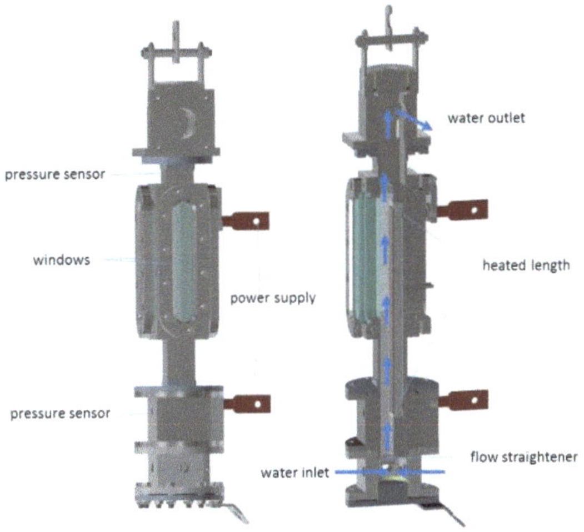

Figure 1 CAD-model of the new test section with planar heater

In cooperation with our project partner KIT/IATF, a new test section with a flat heater (heated length 300 mm, width 20 mm) was developed (Figure 1) for the low pressure loop. The heater is made of stainless steel and extensively instrumented for the measurement of temperature distribution, heating power, pressure gradient and mass flow. In addition, sight glasses from three sides allow the use of optical measurement technology. This test section uses temperature-resistant 3D-printed plastic parts for the first time in COSMOS-L. A flow straightener and several supports for thermocouples at the heater were manufactured. The components can be used permanently at temperatures significantly above 200°C and have already proven themselves in the first experiments.

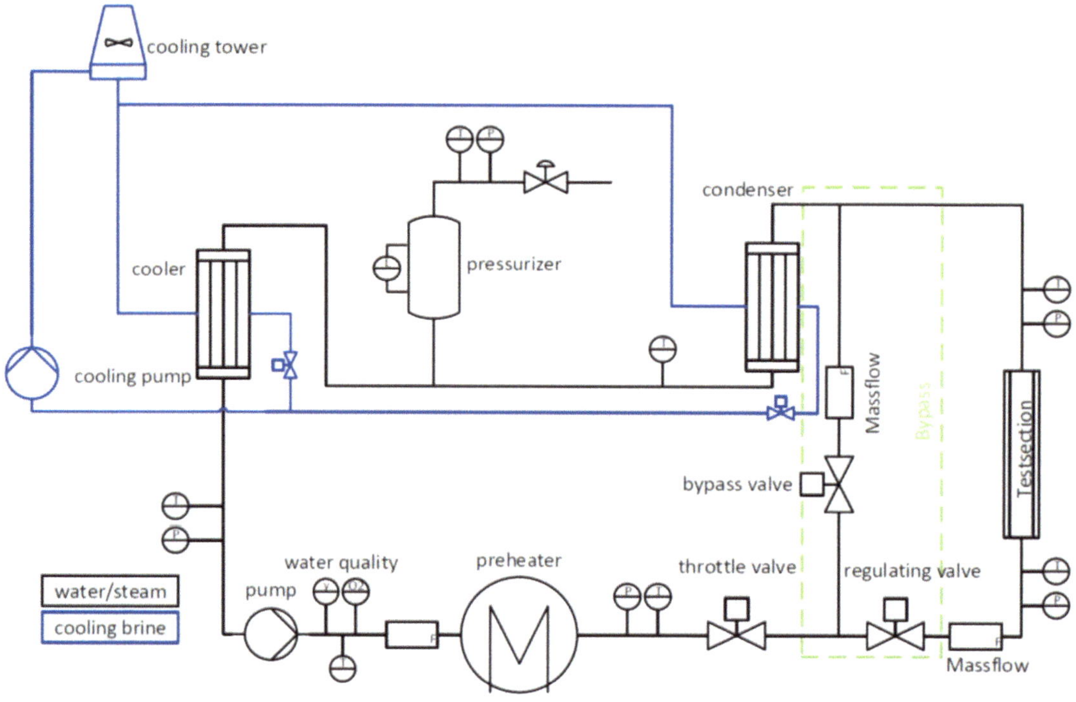

Figure 2 Simplified P&ID of COSMOS-L with new test section bypass

In addition, COSMOS-L was extended with a bypass parallel to the test section (Figure 2). It is controlled by fast-switching proportional valves and has an additional flow meter in order to measure the mass flow through the test section accurately. The mass flow can now be varied via the valves without having to change the total mass flow of the pump. The profile of the variation is thereby programmable. Test track and water loop were put into operation and a first series of tests was carried out. Figure 3 shows two snapshots of the overflowed

Parameter	Value
Pressure	1200 mbar
T_{Inlet}	60°C
Subcooling	45 K
Mass flow density	185 kg/m²s
Heating power	9,3 kW

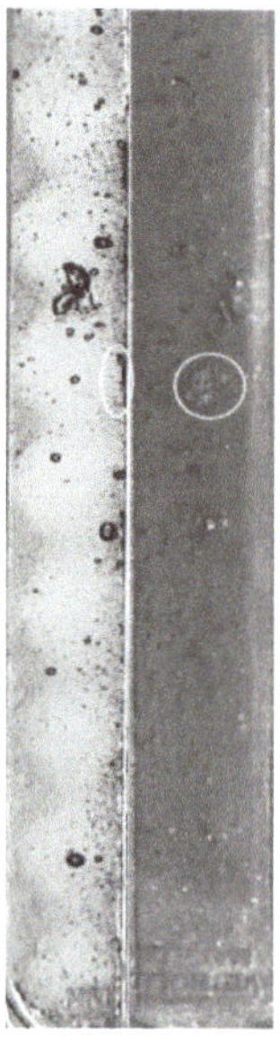

Parameter	Value
Pressure	1200 mbar
T_{Inlet}	60°C
Subcooling	45 K
Mass flow density	185 kg/m²s
Heating power	0 kW

Figure 3 Instant image of the heated section shortly before the boiling crisis (l) and immediately after the power discharge (r).

heater from these tests. Both images show the side view on the left and the front view on the right of the test track respectively.

First, a number of tests were carried out to determine the measurement uncertainty of the critical heat flux. Subsequently, the first parameter variations were conducted. As the main parameter of the investigation, both the period and amplitude of the oscillation were varied. Experiments were started in which the mass flow oscillates periodically (Figure 4).

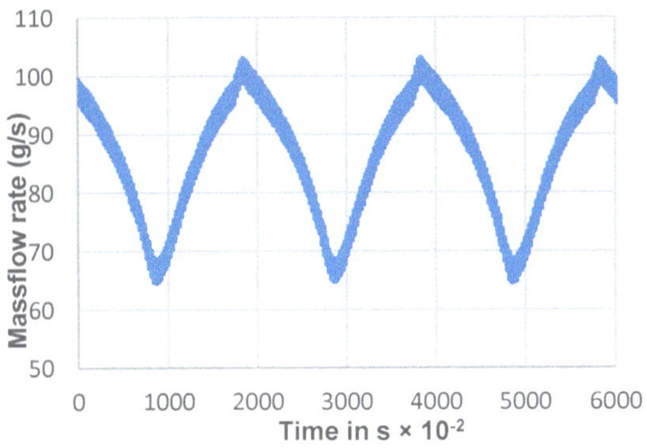

Figure 4 Oscillating mass flow during the first experiments

During the tests, the temperature distribution was observed through thermocouples attached to the back of the heater. The boiling crisis was detected by the associated temperature excursion at the heater and the heating power was dropped to protect the test section (Figure 5).

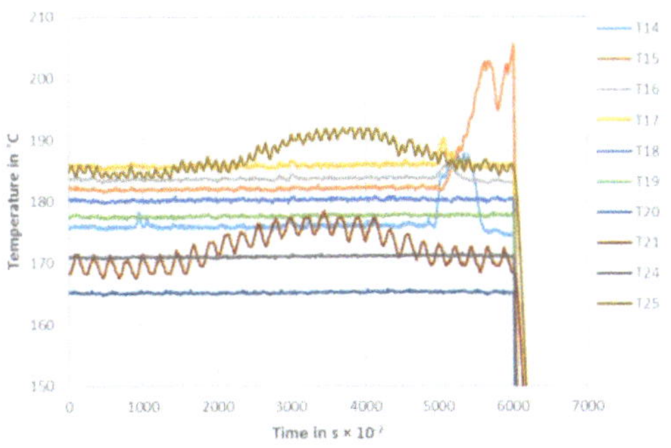

Figure 5 Temperature distribution at the heater

The values obtained this way already show a measurable influence of the fluctuations in the first tests. In Table 1 the example of two selected measuring points shows a reduction of the critical power with fluctuating mass flows of approx. 2.7% with a measurement uncertainty of approx. 1%. This already occurs with a mass flow amplitude of 30% of the nominal value; in the postulated scenario of an IVR, much larger amplitude values can also occur.

Several test series with additional optical instrumentation and parameter variations will be conducted with this test setup over the next few months. In order to investigate larger mass flows with constant inlet subcooling, the thermal power of COSMOS-L facility will be increased in the near future by an additional heat exchanger in order to be able to investigate larger mass flows with constant inlet subcooling.

Exp.	Pressure mbar	Temperature inlet °C	Subcooling K	Massflow-rate kg/m²s	Periode s	Amplitude %	Critical Power kW
1.1	1200	75	45	400	-	-	8.93
1.3	1200	75	45	400	50	30	8.69

Figure 6 Extract from the test evaluation of the first test series. Comparison of a measurement with stationary mass flow and a measurement with fluctuating mass flow

COSMOS-H

Complementary to the activities on the low-pressure facility COSMOS-L, the completion and commissioning of the high-pressure facility COSMOS-H is currently in progress. In 2020, further subsystems of the plant were finalized. Figure 7 shows the simplified flow diagram of the plant, which has two cooling circuits in addition to the high-pressure circuit. With an installed thermal power of approx. 2 MW and a working pressure of up to 17 MPa at approx. 360°C, the water/steam circuit will reach power plant conditions in test operation.

In the EU project McSafer, which started in 2020, investigations into the safety of various SMRs [1] are planned together with the project partner KIT/INR and other partners from research and industry. In COSMOS-H, experiments on heat transfer from nucleate boiling up to the critical heat flow will be carried out on two different test setups, a single tube in the annular gap and a rod bundle consisting of five zircalloy-4 tubes. Comparable experiments have already been performed with a smaller heated length under low pressure conditions at COSMOS-L [2, 3].

Figure 8 shows the first test arrangement. The heated section was constructed and the first components have already been manufactured. The assembly will be fully built up in the following months. In addition to thermocouples and pressure sensors, laser-based glass fibre sensors [4] will also be used in these experiments to measure the phase distribution.

To create a well-defined inflow condition, a flow straightener will be positioned at the inlet of the annular gap. Caused by the electrical, thermal and hydraulic requirements, this part was realized as a ceramic 3D print [Figure 9].

Furthermore, the test track has several sight glasses so that optical methods such as high-speed videometry can also be used in future experiments. As the test track has an increased risk of leaks due to the glass windows, a multi-stage safety concept was created for the plant. For this reason, a fragment shield was constructed around the test section to catch glass splinters and the steam jet in the case of a window failure [Figure 10].

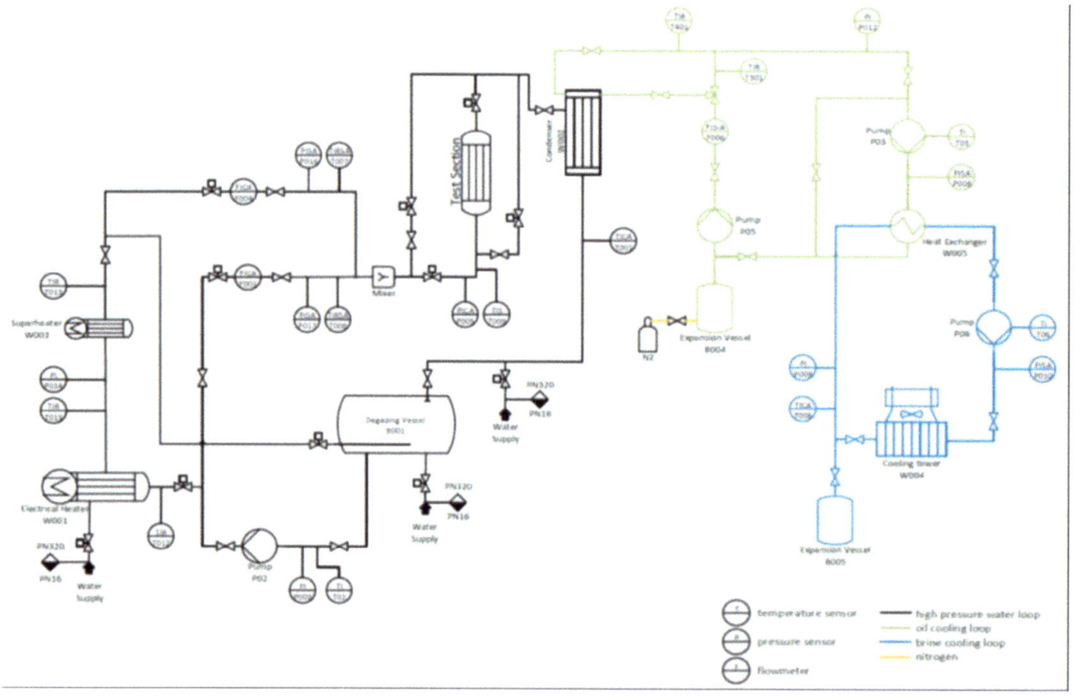

Figure 7 Simplified P&ID of COSMOS-H

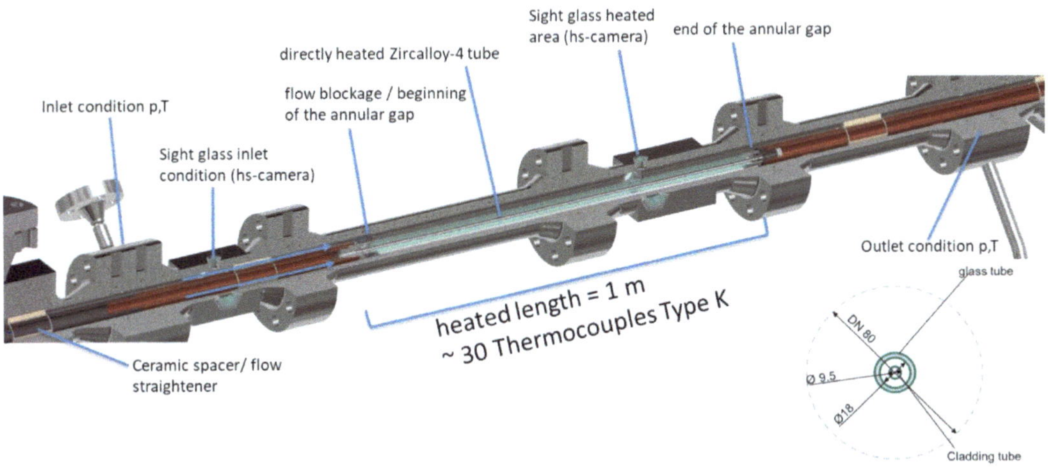

Figure 8 Test arrangement with a single tube in an annular gap

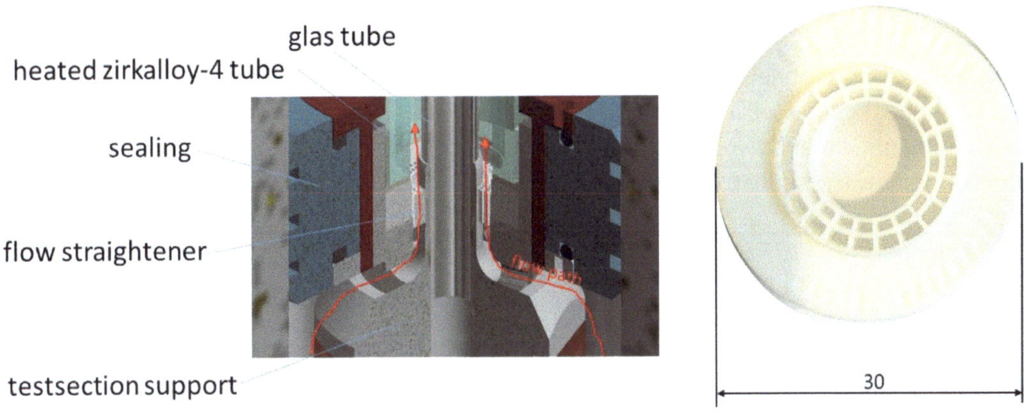

Figure 9 Flow straightener in CAD model and as 3D printed part made of AL_2O_3 ceramics

Figure 10 Pressure hull of the test section including high-pressure windows (l) and fragment shields (r)

Acknowledgements

We would like to thank Prof. Xu Cheng and Stelios Michaelides for the fruitful cooperation in the project KEK-SIMA (FKZ1501577). Furthermore, we would like to thank Victor Sanchez for his support and the good cooperation in Project McSafer (GA No 945063).

References

1. Reitsma, F.; Subki, M.H.; Luque-Gutierrez, J.C.; Bouchet; S.; Advances in Small Modular Reactor Technology Developments, A Supplement to: IAEA Advanced Reactors Information System (ARIS) 2020 Edition; http://aris.iaea.org.

2. Kaiser, F.; Gabriel, S.; Albrecht, G.; (2017); Investigation of the Critical Heat Flux in an Annular Gap and a Rod Bundle Configuration under low Pressure Conditions. ProcessNet-Fachgruppe Wärme- und Stoffübertragung, Bruchsal, 16.-17. Februar 2017

3. Kaiser. F.; Dietrich, B.; Gabriel, S.; Wetzel, T.; Experimentelle Ermittlung von kritischen Wärmestromdichten bei reaktortypischen Bedingungen als Validierungsdaten, Final Report Project NUBEKS (BMWi 1501473B, 27.02.2019)

4. Heineken, F.; Weiterentwicklung von lasergestützten Glasfasersonden zur Messung der Phasenanteile in Zweiphasenströmungen, Bachelorthesis KIT, ITES, 2020

Group: Karlsruhe Liquid metal LAboratory (KALLA)

Liquid metal technology research at Karlsruhe Liquid Metal Laboratory 2020 – Fueling sustainable energy and process technology by innovative approaches

Markus Daubner, Frank Fellmoser, Christoph Hofberger, Marta Kamienowska, Ralf Krumholz, Tim Laube, Karsten Litfin, Franziska Müller-Trefzer, Klarissa Niedermeier, Leonid Stoppel, Leonid Stoppel, Neele Uhlenbruck, Thomas Wetzel, Kurt Wittemann

Introduction

Research and development activities at the Karlsruhe Liquid Metal Laboratory (KALLA) in 2020 can be categorized in three pillars:

- Thermal fluid dynamics of liquid metals
- Liquid metal based heat storage
- Liquid metal based process technology

In all three areas, experimental and modelling approaches are combined, mutually benefitting from each other. All three areas do also benefit from the long lasting liquid metal engineering experience at KALLA and contribute to its progress at the same time. This holds for the design of the loop facilities, reactor setups, instrumentation, components as well as for process control and data acquisition systems. They all define the state of the art in liquid metal technology and enable innovative applications. The latter range from liquid metal based high temperature thermal storage up to highly innovative process routes for e.g. the production of energy carriers.

High temperature heat storage is expected to be an enabling technology for post-fossil heat supply to high temperature thermal process technologies, based on fluctuating sources or waste heat. In terms of high temperature processes itself, liquid metals become more and more visible as heat transfer or even reaction media in highly innovative process routes like direct dehydrogenation of methanol and direct thermal decomposition of methane. While the first can be part of a process chain to produce sustainable liquid energy carriers, the second can be used as hydrogen production technology or as part of innovative pathways towards utilizing carbon dioxide as a raw material for the production of e.g. carbon black. All of the mentioned topics fit together in a picture of enabling technologies and new ideas for the energy transition with its challenges not only for the energy sector, but for chemistry, transport and many others.

In terms of serving fundamental research, a refurbishment of the GALINKA-loop has been accomplished, that will facilitate precise measurements of the heat transfer behaviour in non-homogeneously heated tubes, delivering both – correlations as well as validation data for sophisticated simulation studies and turbulence model development. A second important experimental topic with direct relevance for simulation tool development, validation and safety assessment of innovative nuclear research installations is the investigation of local blockages and deformations in liquid metal cooled rod bundles. Here we continue the traditional combination of generic investigations with application relevant parameter ranges, resulting in unique data sets that are acknowledged worldwide in nuclear safety research.

We are grateful that - despite the difficulties of the pandemic situation - we could continue the strong partnerships with researchers across Europe and worldwide, and even establish new partnerships with internationally leading companies in the fields of liquid metal and process technology. Please see brief descriptions of all mentioned activities on the following pages and do not hesitate to contact us, if you see an opportunity to develop creative ideas with the KALLA team in future.

Thermal Fluid Dynamics of Liquid Metals

GALINKA – Gallium-Indium-Tin loop Karlsruhe

One of the most traditional installations in KALLA, the liquid metal test loop GALINKA, has been completely renewed in order to realize high-precision liquid metal thermohydraulic experiments at room temperature. The used eutectic composition of gallium, indium and tin has a melting point of 10.5°C. The loop is shown on the left-hand side of figure 1. It hosts a variety of instrumentation to determine mass and volume flow rate, temperatures and pressures at several places in the loop. Experiments with heat loads up to 20 kW, which can be cooled back with a cooling water circuit and a liquid metal volume flow rate up to 1.5 m³/h can be realized. Due to the low operating temperature and the low installation space of the loop compared to other liquid metal loops, new experiments can be realized easily and cost effective.

In a current project supported by the German Research Foundation (DFG) (WE 4672/4-1), the influence of an asymmetric thermal boundary condition on the heat transfer in a turbulent liquid metal pipe flow is under investigation. This type of boundary condition can be found in concentrating solar power plants [1, 2]. For this purpose, a test section has been developed to realize either symmetric or asymmetric thermal boundary conditions. Pacio et al. [3] already showed, that even for a symmetric thermal boundary condition high-precision thermohydraulic experiments would be beneficial. Prior to the liquid metal experiments, the new test section has been validated using water at different Prandtl-Numbers and the results were compared to Gnielinski's correlation for heat transfer in turbulent pipe flows [4] showing a good conformity. An according parity plot can be seen figure 1.

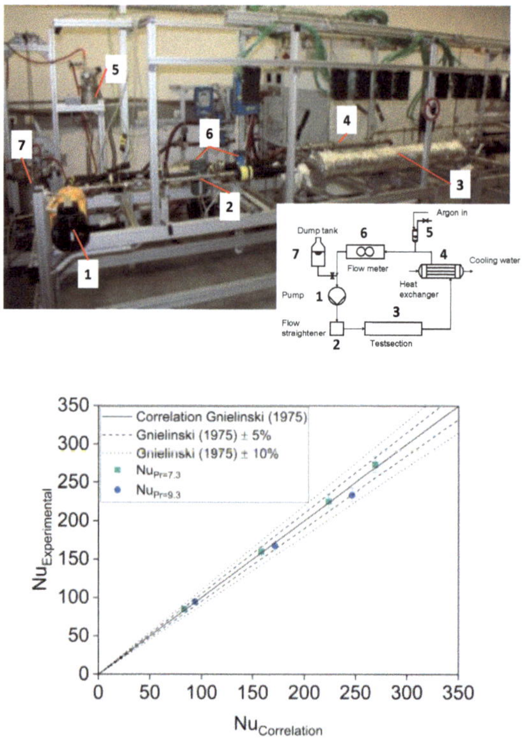

Figure 1: GALINKA loop with all essential components (top), validation of testsection using water experiments and Gnielinski's correlation for heat transfer in turbulent pipe flows (bottom).

References

[1] Marocco, L., Cammi, G., Flesch, J., Wetzel, Th.: Numerical analysis of a solar tower receiver tube operated with liquid metals, International Journal of Thermal Sciences 105 (2016), pp. 22-35

[2] Pacio, J., Singer, Cs., Wetzel, Th., Uhlig, R.: Thermodynamic evaluation of liquid metals as heat transfer fluids in concentrated solar power plants, Applied Thermal Engineering 60 (2013), pp. 295-302

[3] Pacio, J., Marocco, L., Wetzel, Th.: Review of data and correlations for turbulent forced convective heat transfer of liquid metals in pipes, Heat and Mass Transfer 51 (2015), pp. 153-164

[4] Gnielinski, V.: Neue Gleichungen für den Wärme- und den Stoffübergang in turbulent durchströmten Rohren und Kanälen, Forschung im Ingenieurwesen 41 (1975), pp. 8-16

Fuel assembly experiments in Heavy liquid metal flows

For the safe and reliable operation of a fuel assembly in the reactor core, the knowledge of the heat transfer to the coolant is essential. Moreover, during the lifecycle of the assembly its geometry can be deformed by swelling, creeping and mechanical defects or blocked by debris. As a consequence, locally reduced cooling and hot spots are expected.

A set of two different experiments and accompanying CFD simulation is planned in support of safety studies for heavy liquid metal cooled fast reactor systems in the EU-Projects PATRICIA and PASCAL. In a first study, the effect of a well-defined porous blockage in a wire spaced 19-pin rod bundle will be investigated. Following the EU-Project MAXSIMA, where a total blockage has been investigated, the effect of a more realistic sintered blockage with well defined porosity will be investigated in PATRICIA. Detailed instrumentation will give insights about local hot spots as well as recirculation patterns which will be used for the validation of numerical models in house by the Framatome Professional School (FPS) Group of the ITES and external partners in the framework of the EU-project. In a second study, the effect of deformations in a rod bundle on the local flow and temperature field will be studied. In the framework of EU-Project PASCAL, a water rod bundle experiment will be set up at the Karman Institute for Fluid Dynamics (VKI), Brussels with design and numerical support by KALLA.

a)

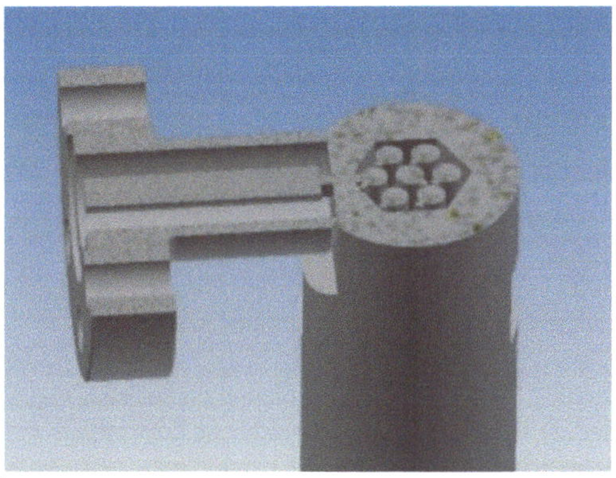

b)

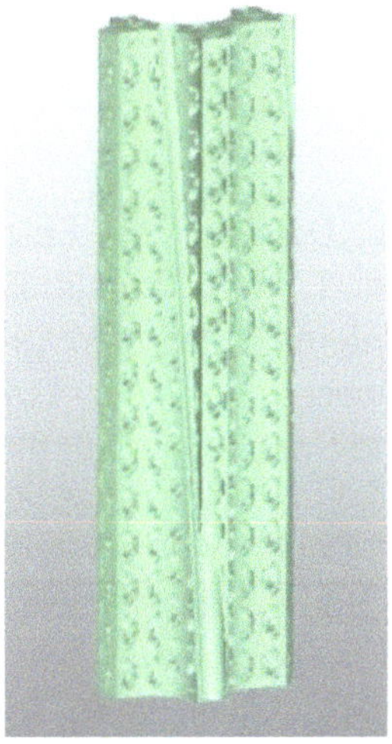

Figure 2: a) Design detail for the planned rod bundle deformation experiment in EU-Project PATRICIA b) Design detail for the planned rod bundle blockage with well-defined porosity in EU-Project PATRICIA

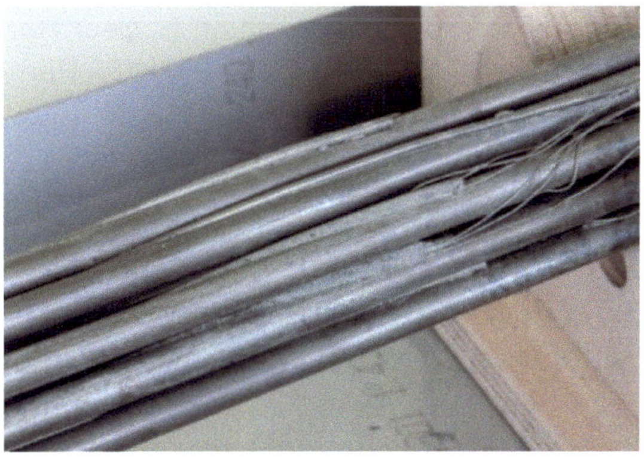

Figure 3: Complete blockage of two sub channels for the experiments in EU-Project MAXSIMA, wires are part of the temperature sensor instrumentation within the blockages and the surrounding rod walls.

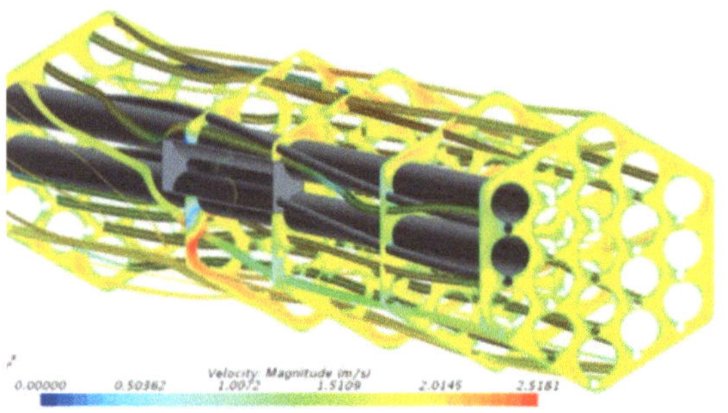

Figure 4: Velocity magnitude plotted on selected planes near the blockage and the streamlines colored with velocity magnitude for the blockage experiments in MAXSIMA[1].

References

[1] Pacio, J.; Daubner, M.; Fellmoser, F.; Litfin, K.; Wetzel, T.; Heat Transfer Experiment in a Partially (Internally) Blocked 19-Rod Bundle with Wire Spacers Cooled by LBE, Nuclear Engineering and Design 330 (2018) 225-240

[2] Pacio, J.; Daubner, M.; Wetzel, T.; Local Blockages in a Rod Bundle with Wire Spacers: Heat Transfer in LBE for the Safety Assessment of MYRRHA, Advances in Thermal Hydraulics (ATH 2018) 2018 ANS Winter Meeting, Orlando, Florida, November 11-15, 2018

[3] Batta, A.; Class, A. G..; Thermalhydraulic CFD Validation for Liquid Metal Cooled 19-Pin Hexagonal Wire Wrapped Rod Bundle, 8th International Topical Meeting on Nuclear Reactor Thermal Hydraulics (NURETH-18), Portland, Oregon, USA, August 18-23, 2019

[4] 1 Batta, A.; Class, A. G.; Pacio, J.; NUMERICAL ANALYSIS OF A LBE-COOLED BLOCKED 19-PIN HEXAGONAL WIRE WRAPPED ROD BUNDLE EXPERIMENT CARRIED OUT AT KIT-KALLA WITHIN EC-FP7 PROJECT MAXSIMA, The 8th European Review Meeting on Severe Accident Research (ERMSAR), Warsaw, Poland, May 16-18, 2017

Liquid Metal Based Heat Storage

Liquid metals are tested for the use in thermal energy storage systems at Karlsruhe Liquid Metal Laboratory (KALLA), as they facilitate highly efficient heat transfer in a wide temperature range. A packed-bed storage system with filler material was proposed as a promising storage system in previous theoretical works at KALLA [1,2]. Starting in 2020, experimental tests are carried out, following a series of subsequent experiments with increasing complexity, size and technology readiness level during the upcoming years. First, the operability of a liquid metal thermal energy storage with a packed-bed has been successfully shown in the lab-scale experiment VESPA [5] with spherical ceramic particles as filler material and lead-bismuth as the heat transfer fluid [3,4]. Figure 5 shows a picture of the open storage container filled with zirconium silicate particles.

The material of the filler material has been selected in cooperation with the group of Prof. Müller and Dr. Weisenburger at the Institute for Pulsed Power and Microwave Technology (IHM) at KIT by storing various candidates (ceramics, glasses and natural stones) in lead-bismuth at very high temperatures in order to examine their corrosion behaviour via optical and electron microscopy and other advanced characterization methods. Important lessons regarding filling, draining, cyclic operation, measuring of temperature and flow have been learned for the planned next experiment, featuring a demonstrator with 100 kWh thermal energy storage capacity that is to be built and tested at KALLA.

Figure 5: Packed-bed thermocline storage in the pre-experiment VESPA at KALLA (Photo: F. Müller-Trefzer)

Liquid metals are especially suitable for high temperature applications, as they are liquid up to temperatures of more than 1000°C. However, their corrosive nature requires technical solutions regarding the structural materials that can be used, especially at temperatures beyond 600°C. In this regard, a BMWi-project together with the German Aerospace Center (DRL) and the industry partner KSB, an international producer of pumps and valves, is starting in 2021 regarding testing of pumps and valves at temperatures up to 700°C [6], again also including the colleagues from the group of Prof. Müller and Dr. Weisenburger at IHM.

References:

[1] Niedermeier, K., Flesch, J., Marocco, L., Wetzel, Th. (2016) Assessment of thermal energy storage options in a sodium-based CSP plant. Applied Thermal Engineering, 107, 386–397.

[2] Laube, T., Marocco, L., Niedermeier, K., Pacio, J., Wetzel, T. (2020) Thermodynamic Analysis of High-Temperature Energy Storage Concepts Based on Liquid Metal Technology. Energy technology, 8 (3)

[3] Niedermeier, K.; Müller-Trefzer, F.; Weisenburger, A.; Wetzel, T. (2020) High-temperature storage with liquid metals - Design of a prototype storage system and material testing. 2nd International Workshop on Carnot Batteries (2020), Online, 15.–16. September 2020

[4] Niedermeier, K.; Müller-Trefzer, F.; Daubner, M.; Marocco, L.; Weisenburger, A.; Wetzel, T. (2020). Theoretical and Experimental Studies of Dual-Media Thermal Energy Storage with Liquid Metal. 26th SolarPACES (2020), Online, 28. September–2. Oktober 2020

[5] Dieterle, K.; Inbetriebnahme und experimentelle Charakterisierung eines Wärmespeichers mit Flüssigmetall als Wärmeträgermedium, M. Sc. Thesis. Karlsruhe Institute of Technology (KIT), 2020

[6] Heidelberger, M. (2021) Erneuerbare Energien: Auf dem Weg zu thermischen Großspeichern. KIT press information 034/2021.

Liquid Metal Based Process Technology

DECAGAS^{LM} - Decarbonisation of nature gas in liquid metal

Hydrogen technology is seen as a key to the success of the energy transition. In June 2020, the German government adopted the national hydrogen strategy [1]. These extensive measures and investment packages are intended to establish hydrogen as a predominant energy carrier in the foreseeable future.

In recent years, intensive work has been carried out at KALLA on a continuously operable process for direct thermal methane pyrolysis as a viable process to produce hydrogen and solid carbon from methane and – perspectively – biogas and natural gas. In the process, the extensive experience with liquid metals, established at KALLA, could be transferred into successful lab-scale experiments, which to date have repeatedly proven the possibility to reach high methane conversion rates and their dependency on operation parameters like temperature [2, 3].

Together with the industrial partner Wintershall Dea [4], the technique is now being further developed for the use on an industrial scale. The basis is the pyrolysis of methane in a liquid metal bubble column reactor (see Fig. 6). In this process, liquid tin is flown through by gaseous methane at 800 °C - 1200 °C. Due to the temperature conditions, the methane is split into two components, solid powdered carbon and gaseous hydrogen. Since the solid carbon has a much lower density than the liquid tin, it accumulates on the surface of the liquid metal and can be removed from the reactor. The gas phase, containing a mixture of hydrogen and unreacted methane, will be further processed according to the requirements of the final application. Unreacted methane and other hydrocarbons can be fed back into the reactor.

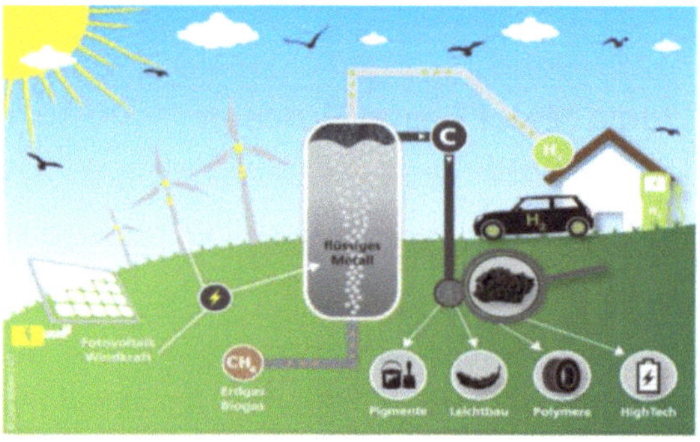

Figure 6: Graphical illustration of methane pyrolysis using bubble column reactor. (Infographic: Leon Kühner, KIT) [4]

References:

[1] Die nationale Wasserstoffstrategie. BMWi, 2020. https://www.bmwi.de/Redaktion/DE/Publikationen/Energie/die-nationale-wasserstoffstrategie.pdf

[2] Geißler, T.: Methan-Pyrolyse in einem Flüssigmetall-Blasensäulenreaktor. Dissertation. Karlsruher Institut für Technologie. 2017.

[3] Stoppel, L.; Fehling, T.; Geißler, T.; Baake, E.; Wetzel, T.: Carbon dioxide free production of hydrogen. In: IOP Conference Series: Materials Science and Engineering 228 (2017), Nr. 1, No. 12016.

[4] Press information 141/2019. Wasserstoff aus Erdgas ohne CO_2-Emissionen. https://www.kit.edu/kit/pi_2019_wasserstoff-aus-erdgas-ohne-co2-emissionen.php

NAMOSYN - Sustainable mobility with synthetic fuels

The BMWi project NAMOSYN (Nachhaltige Mobilität mit synthetischen Kraftstoffen) was created and started in 2019 [1]. Its main aim is to develop and test synthetic fuels that can be produced and used sustainably. One promising possibility is to use formaldehyde as an intermediate to synthesise, e.g. OME, an alternative to diesel. Moreover, it is also by far one of the most commonly produced chemicals in the industry. Nevertheless, a stoichiometric amount of water is formed in the current methods of CH_2O production and needs to be separated, which makes these processes costly.

In contrast, the direct dehydrogenation of methanol with sodium vapour as catalyst may allow to avoid the energy-consuming water separation process and to obtain highly concentrated formaldehyde [2]. Despite these potential advantages, there is no industrially mature method for the production of anhydrous formaldehyde yet. The MEDENA set-up is being constructed at KALLA in order to prove or develop according process concepts [3]. A number of parameters, such as the amount of sodium, the mole fraction of methanol and temperature, play a key role in this type of reaction and will be investigated in detail. The MEDENA set-up aims to demonstrate that vaporised elemental sodium homogenously catalyses the dehydrogenation of methanol to anhydrous formaldehyde with high activity and selectivity.

Figure 7: MEDENA set-up for the dehydrogenation of methanol to anhydrous formaldehyde at KALLA (Photo: M. Kamienowska)

References

[1] Project website: http://namosyn.de/

[2] Ruf, S.; May, A.; and Emig, G.; “Anhydrousformaldehyde by sodium catalysis”, Appl. Catal. A. Gen., vol. 213, no. 2, pp. 203-215, May 2001.

[3] Cortelezzi, D.; Auslegung einer Versuchsanlage zur trockenen Dehydrierung von Methanol zu Formaldehyd, M. Sc. Thesis. Karlsruhe Institute of Technology (KIT), 2020

NECOC – Negative Emissions by Separation of atmospheric CO_2 into technically usable Carbon Black

The BMWi funded project NECOC (Negative Emissions by Separation of atmospheric CO_2 into economically usable Carbon Black) focuses on greenhouse gas reduction and circular economy based on CO_2. In collaboration with the two start-ups Climeworks Germany GmbH and INERATEC GmbH a group of researchers from the Karlsruhe Liquid Metal Laboratory (KALLA) and the Institute of Thermal Process Engineering (TVT) is building an innovative demonstration plant [1]. The objective of the project is to combine three process steps to transform atmospheric CO_2 into economically usable, solid carbon.

Atmospheric CO_2 is seperated from the air via a direct air capture (DAC) facility developed by Climeworks. The second process step, operated by INERATEC, is the catalytic methanation of CO_2 using microstructured reactor concepts in order to overcome the challenge of the high thermal stability of the CO_2 molecule. In a third step the thermal dissociation of methane, also known as methane pyrolysis, takes place in a liquid metal bubble column reactor developped at KALLA. The innovative technology using liquid tin as heat transport medium prevents the pyrolysis reactor from clogging as the solid carbon produced by methane pyrolysis rises to the liquid metal surface as a powder due to its lower density compared to tin.

Besides demonstrating the feasability of the combined process another objective of the project is a closer analysis of the pyrolytic carbon. Therefore, and for the continuous operation of the process, a modification of the pyrolysis reactor allowing continuous carbon recovery on lab scale has been developed and tested during the last months. At the same time construction works for the demonstration plant started several weeks ago after detailed planning. In addition to the BMWi funding, an equipment application was successful at KIT within the framework of the Excellence Initiative (KIT future fields) and as a result a very well equipped gas chromatograph could be procured. This will make it possible to identify by-products in the product gas stream across a broad spectrum.

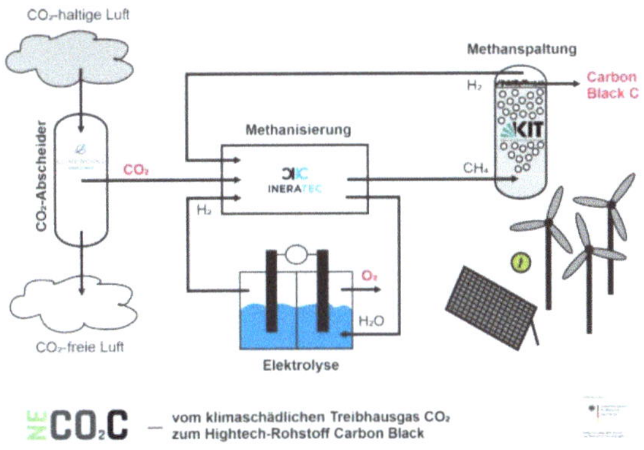

Figure 8: Schematic of the process for converting the greenhouse gas CO2 into technically usable carbon black, which is currently being set up at KIT in a plant

References

[1] Press information 19/2020. Vom Treibhausgas zum Hightech-Rohstoff. https://www.kit.edu/kit/pi_2020_019_vom-treibhausgas-zum-hightech-rohstoff.php

Group: Accident Management Systems

Smart Resilience Engineering: Mitigative Planning of Networked Infrastructures, Uncertainties and Robust Management

S.S. Ottenburger, W. Raskob, T. Müller, S. Möhrle

Introduction

At present, *systemic supply risks* in a future smart urban world, where networking, automation and complexity of infrastructure systems have increased, can hardly be quantified reliably [1]. Part of this uncertainty is that the long-term steady planning and reliable operation of sustainable energy systems, which depend on volatile renewable decentralized feed-in, are based, among other things, on regional climate or weather forecast models. However, the further we look into the future, the more uncertain these are.

For instance, an additional bulk demand for electricity for the use of cooling systems associated with a heatwave [2] can lead to supply bottlenecks and overloads of renewable distribution and transmission grids, ultimately causing large-scale blackouts. The resulting failure cascades in highly networked and automated or electrified systems [3] may lead to considerable supply failures of systemic proportions and significantly reduce the performance of complex *critical infrastructure systems* as typically existing in urban environments. If resilience is not integrated into a system design, renewable energy systems run risk to fail in the long-term. Thus, sustainability and resilience are thought together; this working group develops new methodologies and concepts for identifying resilient planning patterns and smart management systems for adaptive and autonomous energy grids resp., establishes systemic risk assessment frameworks and early-warning concepts, and furthermore designs decision support systems for smart crisis management in complex multi-stakeholder environments.

These research topics belong to the emerging field of 'Smart Resilience Engineering', which leverages lived principles and long-standing expertise of this working group: JRODOS is a decision support system for nuclear emergency management, which has been developed and hosted by this group for many years and is operationalized in more than 40 countries worldwide. However, the gain in knowledge in nuclear emergency management over the last decades is not yet at an end, as on a global level, it can be assumed that nuclear power generation is more likely to be expanded in order to achieve climate targets, and therefore, for example, in the case of Small Modular Reactors (SMRs), which would also be installed close to cities and dense *critical infrastructure systems*, research on the safety of SMRs and emergency protection must be adapted to new siting concepts.

An example of the synergetic effects is the MCDA-KIT[1] [4–6], a general applicable multi criteria decision analysis (MCDA) framework, has been developed by this group in the context of framework, that has been nuclear emergency management (Fig. 1). Originally created for JRODOS, the MCDA-KIT applies MCDA to support decision making in emergency and preparedness, especially if the process is not distinct as measurements and simulation are

[1] https://portal.iket.kit.edu/projects/MCDA/

always affected by uncertainty. On top of that, as in such situations decision-making is in general also supervised by an advisory body, people with different backgrounds and expertise are involved in the process, causing different opinions to collide. Finding a *robust* and acceptable consensus for everyone involved is a difficult task. In this context, MCDA as a supporting method has become increasingly popular. The MCDA-KIT has successfully been applied and enhanced in many European projects [4–6], qualifying the group as advisory for IAEA. Moreover, further collaborations were established, e.g. with the group "Sustainable Bioeconomy" at the Institute for Technology Assessment and Systems Analysis at KIT developing an over-institutional tool, where the main objective is to create a framework for integrated sustainability and resilience assessment.

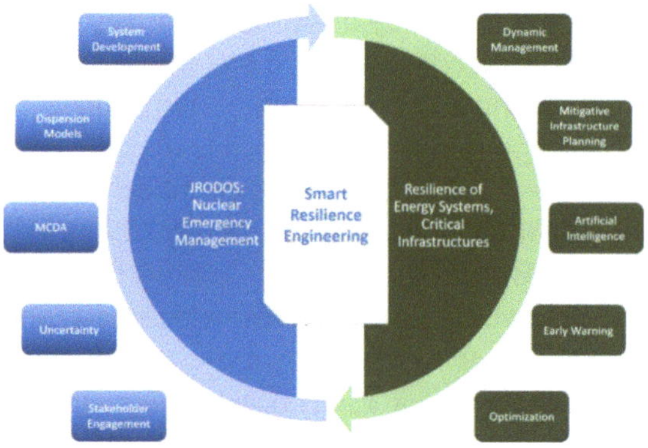

Figure 1 Research in Smart Resilience Engineering develops new methodologies benefiting from long-standing expertise in nuclear emergency management, while in turn adapting modern concepts of resilience engineering leads to innovations in JRODOS; some key competencies of these domains are highlighted in the surrounding boxes.

In conclusion, by combining design questions and management options, this group represents challenging research topics in resilience engineering that deal with different types of uncertainty – thereby, smart methodologies, models, and algorithms are developed, also applying artificial intelligence, in order to identify robust solutions be prepared for unexpected events, to mitigate adverse impacts through resilient designs of networked infrastructures, and to support decision making in effective crisis management (Fig.1).

Resilient Smart Grid Design and Microgrids

Basically, we believe that microgrids (MGs) contribute to a more resilient energy system in view of the aforementioned risks, see e.g. [7]. Applying criticality-based decision criteria for clustering MGs on the distribution level is a promising approach to link critical infrastructure protection to the energy system. Integrating this principle into network planning appears to be a feasible way forward in the context of energy system transformation, in which large investments are expected anyway [8]: this is not about extra measures to strengthen resilience, but about understanding certain topological degrees of freedom in design to increase urban resilience.

In this context, MGs can be seen as built-in redundancies giving rise to a cellular clustering of a distribution area: local resources are able to be self-sufficient in a small area for a certain period of time and furthermore MGs can switch to island mode if the overall grid fails to supply power [9]. How to design MG structures is not cast in stone as well [10–12].

Roughly speaking, the size of a MG depends on the power sources, storages that are installed, the power consumption profiles of the infrastructures/consumers in there, and the maximum expected duration of self-sufficient power provision. Thus, topological degrees of freedom appear in the design of power and smart metering infrastructures but as well in terms of cellular clustering of an energy system.

There are known graph-based network metrics that can be applied as weighted indicators in composite resilience metrics by making use of multi-criteria decision analysis (MCDA), where

Table I: A list of known graph-based network metrics see e.g. [13]

l	characteristic path-length	Efficiency- average of the shortest path lengths
c_b	central point dominance	Dominance of particular nodes
f_c	critical ratio of defragmentation	Removal of a fraction $\geq f_c$ leads to a defragmentation of the network in different clusters – robustness of a network against catastrophic natural events (e.g. earthquake)
λ_2	algebraic connectivity	Expresses the number of disjoint paths, i.e. the network remains fully connected despite the removal of nodes
r_m	meshdness coefficient	Robustness and path redundancy – failure of nodes

the weights can be specified through an analytical hierarchical process (AHP) with the help of stakeholder engagement [13].

If we consider MG clustering, statistical peak load values and initial criticality of network nodes as further attributes of a distribution network, there is the following metric analogously defined as in [14]:

Let $\mathcal{M}$ be the set of all MGs in a smart grid (SG) and $\mathcal{I}_A$ be the index set of all infrastructures that lie in an MG $A \in \mathcal{M}$. Furthermore, we denote the statistical power peak demand of a consumer $i \in \mathcal{I}$ with d_i. A measure for criticality-demand concentration is given by

$$CD_A^{x,y} := \sum_{j \in \mathcal{I}_A} \left(\frac{c_j}{\sum_{k \in \mathcal{I}} c_k}\right)^{1-x} * \left(\frac{d_j}{\sum_{k \in \mathcal{I}} d_k}\right)^{1-y} \quad (1)$$

where $0 \leq x, y \leq 1$ and $x + y = 1$. Depending on the values of x and y, the greater this indicator is the more critical and tense the supply of infrastructures in A and the provision with critical services might get in times of disturbed power supply. The new metric associated to a SG is given as

$$CD := \max_{A \in \mathcal{M}} CD_A \,, \quad (2)$$

where x and y where omitted for better readability.

Hence, CD can be integrated into an MCDA-based resilience metric as described above for urban resilience assessments of smart distribution grids. It is important to point out, that the list of metrics given here does not refer to power physical constraints, however they play a crucial role from a network perspective. Nevertheless, power physical constraints or metrics should be considered to better assess the feasibility of power system designs – this is subject to current research. Furthermore, this approach can be adapted to any other networked infrastructure including hydrogen supply systems.

Resilient Energy Management

The IEEE standard 1366 [15] specifies a series of twelve reliability indices for power distribution systems, the most familiar of which are SAIDI (System Average Interruption Duration

Index) and SAIFI (System Average Interruption Frequency Index). Roughly spoken, SAIDI and SAIFI refer to periods without power supply and are useful in a posteriori assessments of the reliability of power distribution systems. LOLE (Loss of Load Expectation) refers to the number of hours per year during which supply is statistically expected not to meet demand in the long run. Since these indices are determined statistically, they are not subtle enough for the objective of having fine resolved measurements of security of supply as a basis for short-term or real-time decision-making.

The purpose of this subsection is to propose a (global) criticality-based metric that might be used in some sort of composite resilience metrics also taking efficiency and fairness into account [16].

We assume that infrastructures are equipped with smart meter, which are able to prevent power consumption above a specific threshold and furthermore have a certain power demand flexibility that depends on their sub-processes and the current demand of the infrastructure's functions. If $I := \{1, \dots, N\}$ is the set of all infrastructures from an urban distribution network, then we express the demand flexibility of an infrastructure $l \in I$ with the interval $[P^l_{D,min}, P^l_{D,max}]$ and its global criticality with c_l. Of course, there are infrastructures that need to maintain all sub-processes in order to fulfill their main tasks sufficiently, e.g. in dialysis clinics, dialysis machines, pumps, hot water disinfection systems etc. have to be in operation mode concurrently.

Let us consider k intervals of fixed length, e.g. 30 minutes, and let $\mathfrak{sp}_{l,t}$ be the assigned power to infrastructure l in time interval t, which is considered as the power threshold l cannot exceed. We define the following urban resilience indicator for power distribution:

$$\mathfrak{si} := \sum_{t=1}^{k} \sum_{l=1}^{N} \widetilde{c_l} * Q_{l,t}\,, \qquad (3)$$

where – assuming there is at least one $i \in \{1, \dots, N\}$ with $c_i \neq 0$ such that $\widetilde{c_l} := \frac{c_l}{\sum_{j=1}^{N} c_j}$ is well-defined – $\widetilde{c_l}$ is called the normalised variant of c_l and

$$Q_{l,t} := \begin{cases} 0\,, \mathfrak{sp}_{l,t} < P^l_{D,min} \\ \frac{\mathfrak{sp}_{l,t}}{P^l_{D,max}}\,, P^l_{D,min} \leq \mathfrak{sp}_{l,t} \leq P^l_{D,max} \\ 1, P^l_{D,max} < \mathfrak{sp}_{l,t} \end{cases} \qquad (4)$$

characterizes the quality of power supply from the customer's perspective. In [16] first simulation results were presented, which have shown that instead of having controlled load shedding as described in Fig. 3, operationalizing this type of indicator in SGs can result in urban resilient – fair and efficient – power flows avoiding large-scale blackouts and in a significantly improved $\mathfrak{si}$, see Fig. 2.

Instead of having a controlled load shedding as described in Fig. 3, where large areas were switched off in a round-robin manner, smarter distribution could avoid such blackouts and assign power thresholds in an urban resilient and physically feasible way to the customers – i.e. no large-scale blackouts would occur, see Fig. 2.

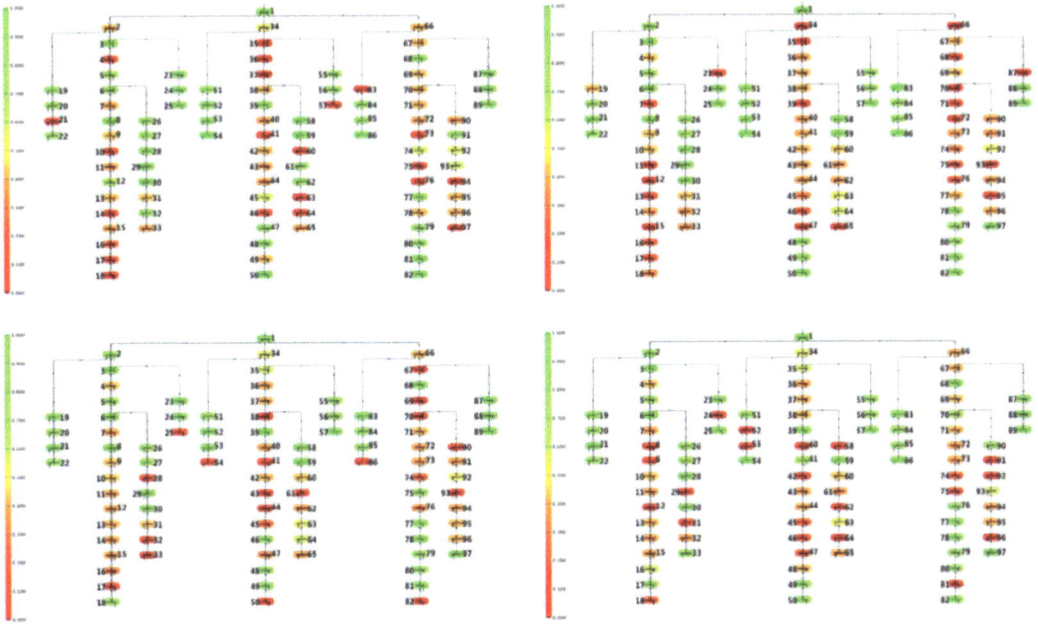

Figure 2 Heat map of the relative power supplies: in contrast to the rolling blackout of Fig. 3 continuous power flows are realised in an urban resilient, fair and efficient way over four time steps (red colour indicates induced blackout) [16]

	Referring to time slots of length 30, 90 minutes ...								
	1	2	3	4	5	6	7	8	9	...
A	●			●			●			
B		●			●			●		
C			●			●			●	

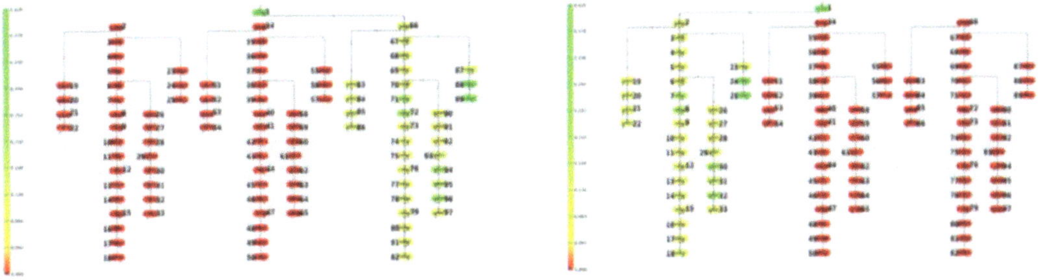

Figure 3 Three supply areas alternately taken off the grid in a round-robin manner – red colour indicates induced blackout [16]

Dynamic Measure Development – Agent-based optimization

Well-known features of agent-based modelling include scalability and simple adaptability of specific agent models and parameters. Besides a flexible integration of new agents, also further *preferences* can be included on an agent-specific level and calibrated or dynamically weighted depending on the agent's environment and forecasts w.r.t. various aspects related to the demands and resource availability e.g. weather forecasts in conjunction with renewable energy generation. Furthermore, based on the various needs of agents, representing different stakeholders or entities in a system, their preferences, and *global strategic goals,* solutions on how to distribute resources in a resilient way can be decided by negotiation- or optimization procedures targeting at identifying feasible and robust measures, and simulated. Agents' preferences and global strategic goals are considered as a result of a two-stage MCDA-process, which might be influenced by boundary conditions or sudden state changes.

Therefore, a resource independent agent-based optimization framework (ABO-F) has been set up that distinguishes between three different agent types: networked-, non-networked-, and purely demanding agents. Once this ABO-F has been instantiated according to a specific context, we speak of a multi-agent system (MAS). This MAS is equipped with a certain resource taxonomy that defines a universal language, which allows agents to communicate particular resource needs as well as offers. First simulation and optimization results were produced in the context of pharmacies (further agent models are to be included [17]) and urban crisis management, see Fig. 4 – as ABO-F is considered as an engine for smart crisis management it is improved and extended continuously, where a particular focus lies on optimization and artificial intelligence. It is envisaged that ABO-F will be integrated into JRODOS.

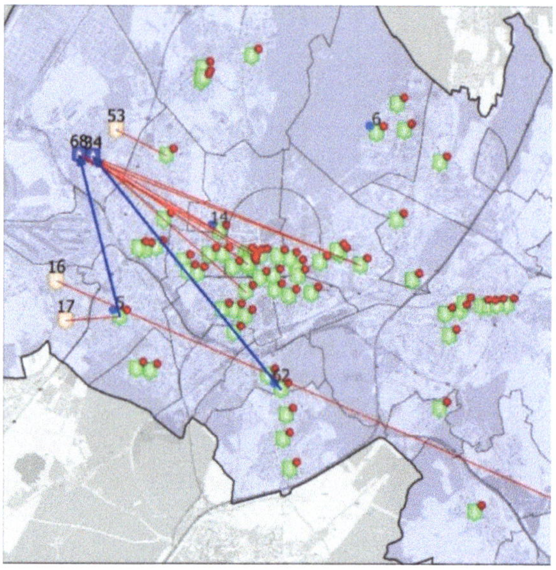

Figure 4 Pharmacies lacking power and medicine have blue icons, pharmacies lacking only medicine have yellow icons. All pharmacies offering medicine or an emergency power unit (EPU) have a red or blue dot resp. The edges describe a possible solution of allocating medicine and EPU.

Nuclear Preparedness and Response

Nuclear preparedness and response remain an important research topic due to research needs identified at the international level within the NERIS[2] (European Platform on preparedness for nuclear and radiological emergency response and recovery) SRA (Strategic Research Agenda) and at the national level with the many Nuclear Power Plants surrounding Germany[3]. To support decision making KIT has coordinated and is coordinating all research activities related to the decision support system JRODOS[4] (JAVA based Real-time Decision Support). The system is able to support decisions about the introduction of a wide range of potentially useful countermeasures

[2] https://eu-neris.net/library/sra.html

[3] https://www.bmu.de/en/topics/nuclear-safety-radiological-protection/nuclear-safety/response-to-fukushima/overview/

[4] https://resy5.iket.kit.edu/JRODOS/

(e.g., sheltering and evacuation of people, distribution of iodine tablets, food restrictions, agricultural countermeasures, relocation, decontamination, restoration, etc.) mitigating the consequences of an accident with respect to health, the environment, and the economy. It can be applied to accidental releases into the atmosphere and into various aquatic environments. Appropriate interfaces exist with local and national radiological monitoring data, meteorological measurements and forecasts, and for adaptation to local, regional and national conditions worldwide.

Nuclear emergencies, as demonstrated with the Chernobyl and Fukushima disasters, do not stop at national borders. However, so far, no harmonized response was established in Europe or worldwide. To overcome this constraint, The European Commission supports the installation of JRODOS in national emergency centres. Having one system operational, at least results are comparable, even if national regulations may differ.

KIT has installed JRODOS in many countries, e.g. starting in 2020 in 6 West-Balkan countries, 6 countries of the Gulf Cooperation Council, Armenia and Iran.

Ongoing research to improve JRODOS is related to source term reconstruction [29] and ensemble modeling to capture uncertainties in the early phase of an emergency [27, 28].

Ensemble Modelling and Uncertainty

The potential source term as well as meteorological forecasts are the main uncertainties in the early phase when the release to the atmosphere is expected or ongoing [26]. On the other hand, evacuation of the population is most effective before the release has started. To investigate these uncertainties, research was conducted within CONFIDENCE[5] and JRODOS has been expanded with ensemble modelling functionalities [28]. One of the questions

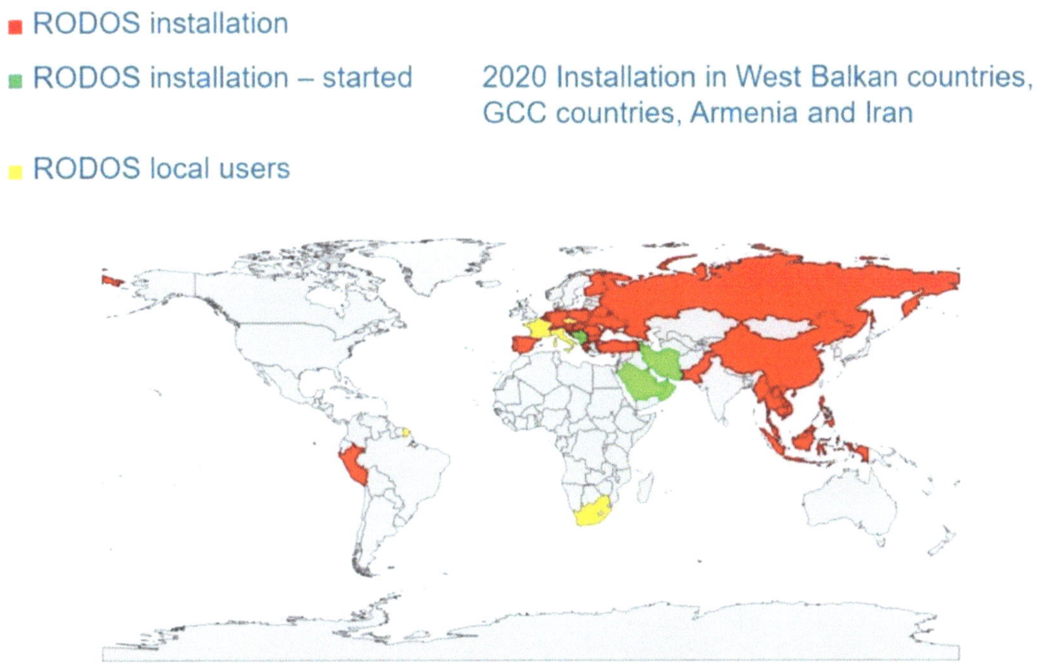

Figure 5 JRODOS installations worldwide.

[5] https://portal.iket.kit.edu/CONFIDENCE/index.php

that was raised but not answered was "do ensembles capture the full uncertainty of a meteorological forecast?".

To look into this question, KIT collected ensembles from three different weather providers

- NOMADS[6] with 20 ensembles
- Canadian[7] weather service with 21 ensembles
- German[8] weather service with 41 ensembles

at distinct times in March 2020. Sites all over the world were selected and ensemble calculations have been performed. To investigate only the meteorological uncertainties, the source term was fixed. The results are presented as percentiles, exceeding a particular dose value. Dark colors mean that many ensemble members fit therein, whereas lighter colors show areas where only some ensemble members predict exceedance of the threshold. The following figures show example results for a more constant meteorological situation and for one with changing wind directions.

Even if preliminary, these two examples indicate that using one provider does not capture the full uncertainty that exist when considering other weather providers. Fig. 7 shows very similar results from all three weather providers; however, the meteorological conditions were quite stable. Fig. 6 on the other hand demonstrates that all three providers differ considerably, at least for the lower percentiles. Nevertheless, Fig. 6 also demonstrates the added value of an ensemble assessment compared to a single forecast. As ensemble modelling is very time consuming – each ensemble member run is one project in JRODOS – a further research question for the following years is the number of ensemble members that is sufficient for operational use, considering the time constraints but reflecting still the existing uncertainty band.

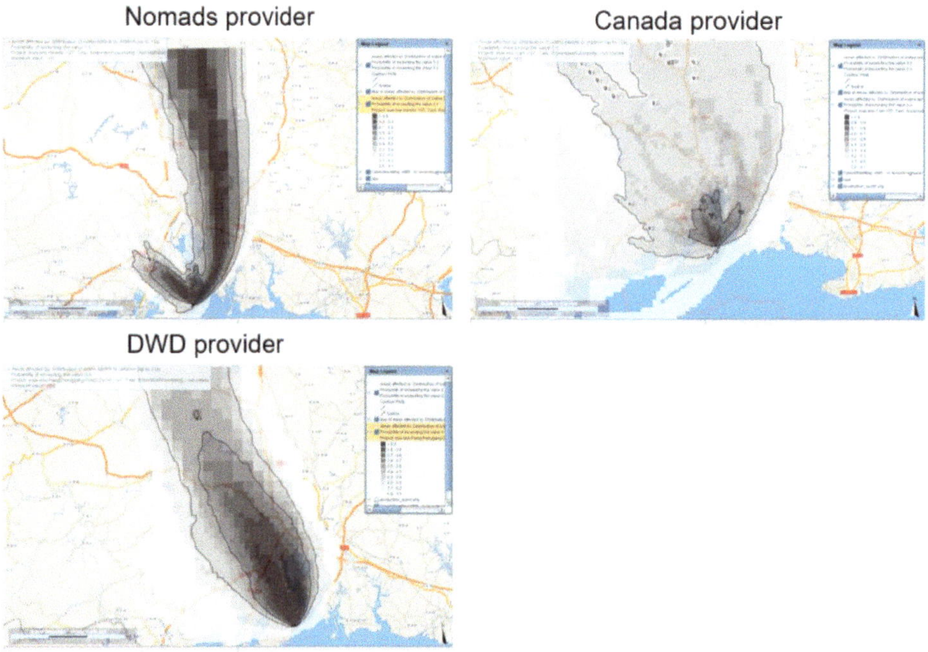

Figure 6 Release in China, exceedance of iodine prophylaxis for children (50 mSv).

[6] https://nomads.ncep.noaa.gov/
[7] https://weather.gc.ca/canada_e.html
[8] https://www.dwd.de/EN/Home/home_node.html

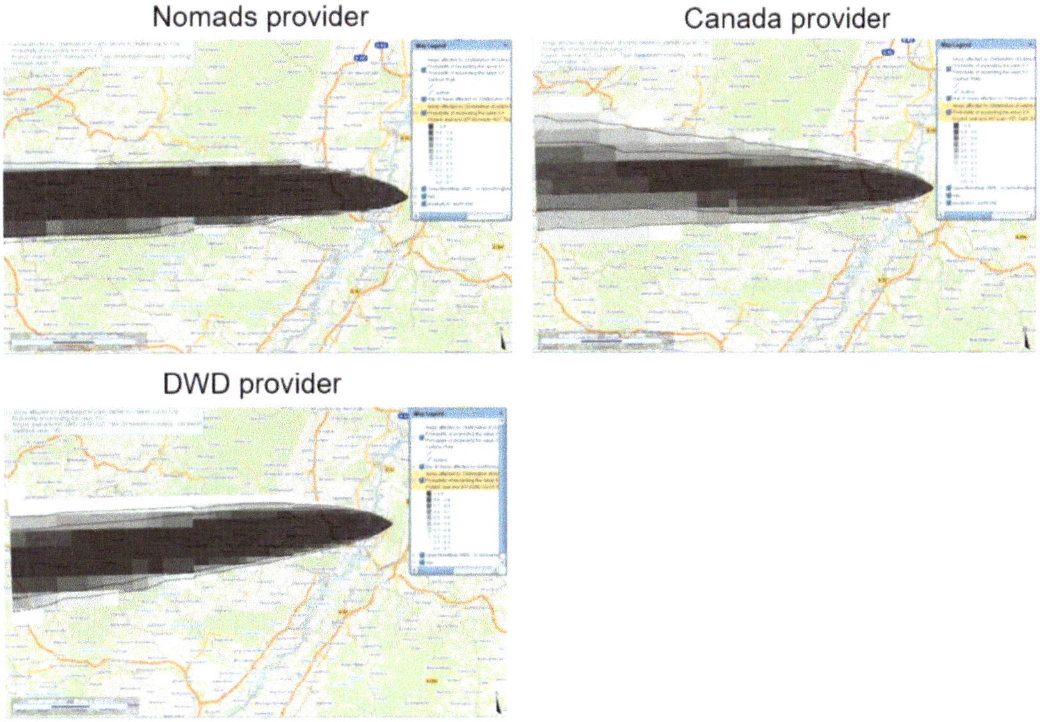

Figure 7 Release in Germany, exceedance of iodine prophylaxis for children (50 mSv).

References

[1] Helbing, D.; "Globally networked risks and how to respond," *Nature*, vol. 497, no. 7447, pp. 51–59, 2013, doi: 10.1038/nature12047.

[2] Mancarella P., *et al.,* "Impact of climate change on the resilience of the UK power system," in *IET International Conference on Resilience of Transmission and Distribution Networks (RTDN) 2015*, Birmingham, UK, Sep. 2015, 8 (6 .)-8 (6 .).

[3] Buldyrev, S. V.; Parshani, R.; Paul, G.; Stanley, H. E. and Havlin, S.; "Catastrophic cascade of failures in interdependent networks," *Nature*, vol. 464, no. 7291, pp. 1025–1028, 2010, doi: 10.1038/nature08932.

[4] Müller, T.; Bai, S. and Raskob, W.; "MCDA handling uncertainties," *Radioprotection*, vol. 55, no. 5, S181-S185, 2020, doi: 10.1051/radiopro/2020030.

[5] Duranova, T.; van Asselt, E.; Müller, T.; Bohunova, J.; Twenhöfel, C.J.W. and Smetsers, R.C.G.M.; "MCDA stakeholder workshops," *Radioprotection*, vol. 55, HS2, S193-S196, 2020, doi: 10.1051/radiopro/2020032.

[6] Duranova, T.; *et al.,* "CONFIDENCE dissemination meeting: Summary on the scenario-based workshop," *Radioprotection*, vol. 55, 5 Suppl, S17-S37, 2020, doi: 10.1051/radiopro/2020009.

[7] Strbac, G.; Hatziargyriou, N.; Lopes, J. P.; Moreira, C.; Dimeas, A. and Papadaskalopoulos, D.; "Microgrids: Enhancing the Resilience of the European Megagrid," *IEEE Power and Energy Mag.*, vol. 13, no. 3, pp. 35–43, 2015, doi: 10.1109/MPE.2015.2397336.

[8] European Parliament, *European Energy Industry Investments.* [Online]. Available: https://www.eesc.europa.eu/sites/default/files/files/energy_investment.pdf (accessed: Jun. 23 2020).

[9] Strickland, E.; *A Microgrid That Wouldn't Quit.* [Online]. Available: https://spectrum.ieee.org/energy/the-smarter-grid/a-microgrid-that-wouldnt-quit (accessed: Jun. 26 2020).

[10] Ottenburger, S. S.; Münzberg, T. and Strittmatter, M.; "Smart Grid Topologies Paving the Way for an Urban Resilient Continuity Management," *International Journal of Information Systems for Crisis Response and Management*, vol. 9, no. 4, pp. 1–22, 2017, doi: 10.4018/IJISCRAM.2017100101.

[11] Cox, W. and Considine, T.; "Structured energy: Microgrids and autonomous transactive operation," in *IEEE PES Innovative Smart Grid Technologies Conference (ISGT), 2013: 24 - 27 Feb. 2013, Washington, DC*, Washington, DC, 2013, pp. 1–6.

[12] Melike, E. K.; Burak, K. and Hussein, M.; "Reliable overlay topology design for the smart microgrid network," *IEEE Network*, vol. 25, no. 5, pp. 38–43, 2011, doi: 10.1109/MNET.2011.6033034.

[13] Pandit, A. and Crittenden, J. C.; "Index of network resilience for urban water distribution systems," *IJCIS*, vol. 12, 1/2, p. 120, 2016, doi: 10.1504/IJCIS.2016.075865.

[14] Ottenburger, S. S.; Bai, S. and Raskob, W.; "MCDA-based Genetic Algorithms for Developing Disaster Resilient Designs of Critical Supply Networks," in *2019 International Conference on Information and Communication Technologies for Disaster Management (ICT-DM)*, Paris, France, Dec. 2019 - Dec. 2019, pp. 1–4.

[15] *IEEE Guide for Electric Power Distribution Reliability Indices*, Piscataway, NJ, USA.

[16] Ottenburger, S. S.; *et al.*, "A Novel Optimization Method for Urban Resilient and Fair Power Distribution Preventing Critical Network States," *International Journal of Critical Infrastructure Protection*, p. 100354, 2020, doi: 10.1016/j.ijcip.2020.100354.

[17] Wandler, S.; Bai, S. and Raskob, W.; "Energy consumption models in dialysis clinics for agent-based decision support," *IJSPM*, vol. 15, 1/2, p. 170, 2020, doi: 10.1504/IJSPM.2020.106981.

[18] Bai, S.; "Agent based model in risk management: several applications," *The Second International Symposium on Simulation and Process Modelling (ISSPM 2020), Shenyang, China, 29.–30*, 2020.

[19] Bai, S. and Raskob, W.; "Agent-based Negotiation Simulation.," *5th NERIS Workshop "Key challenges in the preparedness, response and recovery phase of a nuclear or radiological emergency". Proceedings, 53–62*, 2020.

[20] Bai, S.; Raskob, W. and Müller, T.; "Agent based model," *Radioprotection*, vol. 55, HS1, S187-S191, 2020, doi: 10.1051/radiopro/2020031.

[21] Bai, S.; "Simulations of COVID-19 spread by spatial agent-based model and ordinary differential equations," *IJSPM*, vol. 15, no. 3, p. 268, 2020, doi: 10.1504/IJSPM.2020.10029411.

[22] Bolleman, J. T.; et al., "Semantic Variation Graphs: Ontologies for Pangenome Graphs.," International Conference on Intelligent Systems for Molecular Biology (2020), Online, 13.–16. July,2020, July 16th, 2020.

[23] Duranova, T.; Beresford, N. A.; Perko, T. and Raskob, W.; "Education and training activities in the Euratom CONFIDENCE project," Radioprotection, vol. 55, HS2, S45-S50, 2020, doi: 10.1051/radiopro/2020011.

[24] Möhrle, S.; "Case-Based Decision Support for Disaster Management," 2020.

[25] Nagy, A.; Perko, T.; Müller, T.; Raskob, W. and Benighaus, L.; "Uncertainty visualization using maps for nuclear and radiological emergencies," Radioprotection, vol. 55, no. 1,

S197-S203, 2020, doi: 10.1051/radiopro/2020033.

[26] Raskob, W.; et al., "Confidence: achievements and way forward," Radioprotection, vol. 55, no. 5, S39-S43, 2020, doi: 10.1051/radiopro/2020010.

[27] Raskob, W.; et al., "CONFIDENCE: project description and main results," Radioprotection, vol. 55, 5 Suppl, S7-S15, 2020, doi: 10.1051/radiopro/2020008.

[28] Raskob, W. and Duranova, T.; "Editorial: the main results of the European CONFIDENCE project," Radioprotection, vol. 55, S1-S5, 2020, doi: 10.1051/radiopro/2020007.

[29] Raskob, W.; Haller, C.; Hasemann, I.; Schichtel, T. and Trybushnyi, D.; "Source term reconstruction module in JRODOS.," 5th NERIS Workshop "Key challenges in the preparedness, response and recovery phase of a nuclear or radiological emergency". Proceedings, 112–118, 2020.

[30] Walsh, L.; Ulanowski, A.; Kaiser, J. C.; Woda, C. and Raskob, W.; "A new European cancer risk assessment tool for application after nuclear accidents," Radioprotection, vol. 55, S95-S99, 2020, doi: 10.1051/radiopro/2020018.

[31] Yokoyama, T. T.; *et al.,* "Semantic Variation Graphs: Ontologies for Pangenome Graphs.," *International Conference on Intelligent Systems for Molecular Biology (2020), Online, 13.–16. July,2020, July 15th*, 2020.

Group: Hydrogen

Fundamental Behaviour of Hydrogen to Applied Accident Consequence Analysis for Hydrogen as a Safe Energy Carrier

Abhiram Chavali, Andrei Denkevits, Guang Hu, Olaf Jedicke, Thomas Jordan, Birgit Kaup, Alexei Kotchourko, Mike Kuznetsov, Alexander Lelyakin, Frank Prestel, Maria Stassen, Jianjun Xiao, Zhanjie Xu

Introduction

The Hydrogen Group continued to develop models and in-house specialized reactive CFD codes (GASFLOW and COM3D) and conducted a huge experimental program to deepen the understanding of the mixing behavior and the transient combustion phenomena, flame acceleration and deflagration-detonation-transition, of accidentally released hydrogen. These activities were focussed on phenomena related to cryogenic hydrogen in 2020.

Members of the group are actively transferring their insights and expertise into the respective IEA Hydrogen Task 37, the standards committees of ISO (TC 197), CEN/CENELEC (TC 6) and in the European Hydrogen Safety Panel. Among other third party funded projects the Hydrogen Group has been conducting ambitious experimental programs for the European Fuel Cell and Hydrogen Joint Undertaking (FCH JU) funded projects PRESLHY - Pre-normative research for the safe use of liquid hydrogen - and HyTunnel-CS - Pre-normative research for safety of hydrogen driven vehicles and transport through tunnels and similar confined spaces.

This report present some highlights from the experimental campaigns with liquid, cryogenic respectively, hydrogen conducted in the frame of the European PRESLHY project. The experiments will be presented with increasing confinement of the reactive mixtures and start with a description of pool experiments.

Liqud Hydrogen Pool Experiments

The experimental setup for the Pro Science (PS)/KIT POOL experiments is shown in Figure 1 below. The facility comprises mainly of an insulated square stainless-steel box (50 cm x 50 cm, height: 20 cm) on a scale. Three identical boxes were made and filled up to half the height (10 cm) with different substrates (concrete, sand, water and gravel) prior to the experiments. The concrete pool was prepared already more than one month prior to the experiments, while the experiments with water and gravel were performed with the same box, which was freshly filled with the respective substrate for the corresponding experiment.

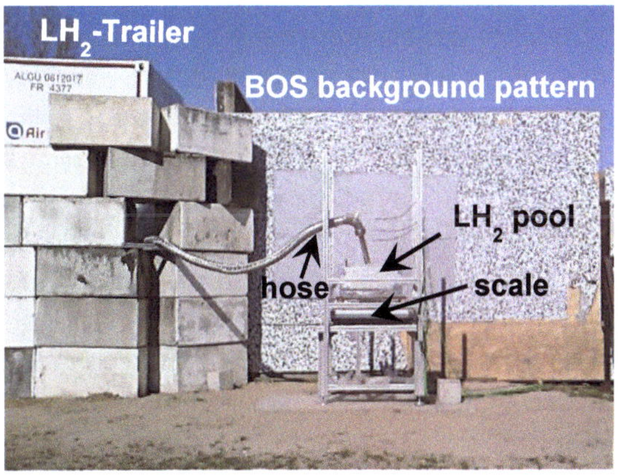

Figure 1: POOL-facility set-up for the unignited and ignited LH2 pool experiments of the PRESLHY project.

Inside and above the substrate, as well as above the facility in total 24 thermocouples were distributed. Above the facility also three measurement positions for remote H2-concentration measurements were placed, see Fig. 2.

Further instrumentation included optical observation and ambient wind measurements. In four of the ten experiments also a large fan was installed to investigate the influence of side wind on the formation and evaporation of LH2-pools.

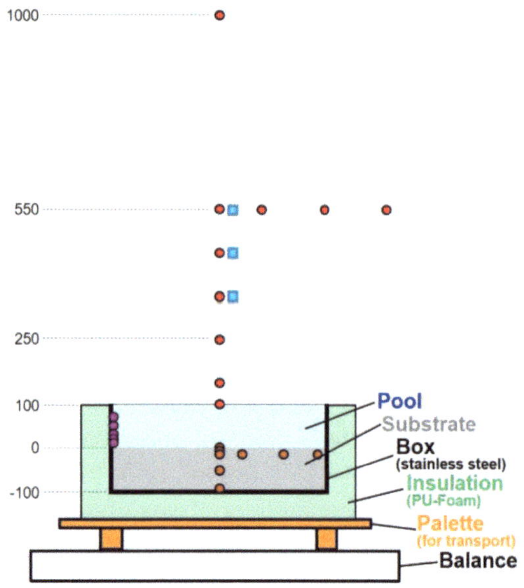

Figure 2: Sketch of POOL-facility set-up

In most of the experiments the pool was filled three times with LH2 until it started to overflow to investigate the influence of different initial temperatures of the substrate. After the fillings the pool was left to evaporate. The filling level of the pool was determined using the weight of the facility and the thermocouples inside the pool in different heights, which clearly indicated their coverage with LH2 by a constant value of approx. 20 K, while they started to show higher transient values as soon as they were exposed to the atmosphere, see Fig. 3.

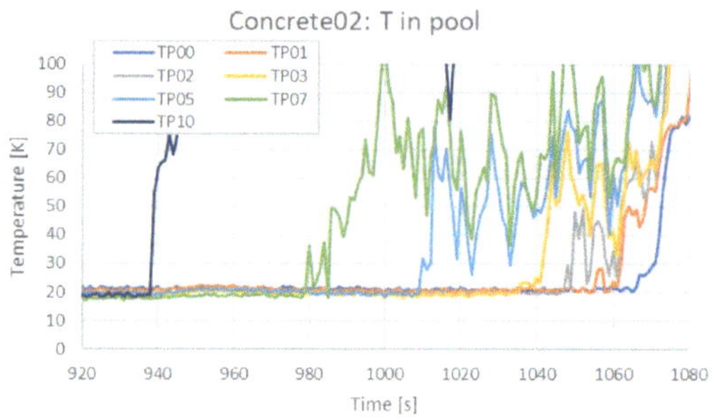

Figure 3: Temperature recordings for experiment Concrete02.

With these measurements the evaporation rates for the different substrates have been determined, see Figure 4. Also the influence of side wind on this phenomenon was assessed. In all cases much faster evaporation rates were determined for side wind conditions (vWind ≈ 5 m/s). A further important outcome of the experiments was information on the concentration distribution above the pools that was used to define the ignition position and point in time for a second experimental series in which the gas cloud above the pools is ignited.

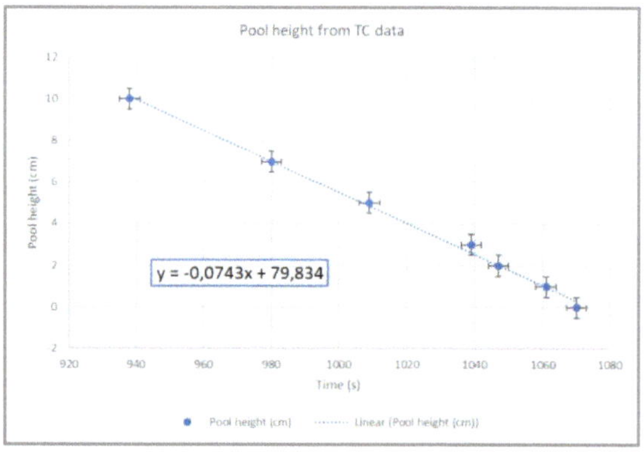

Figure 4: Example for determination of LH2-evaporation rate; here experiment Concrete02.

With the same set-up and experimental procedure also experiments on the ignition of LH2-spills were performed. For these experiments the instrumentation above the pool was removed (compare sketch in Figure 2) and replaced by 6 fast pressure sensors that were positioned in heavy adapters on the ground in the close vicinity of the facility. The gas cloud above the pool was ignited using two electrodes in between which a high frequency spark (60 kV, 200 Hz) was generated. In total 14 ignited POOL-experiments with the four substrates concrete, sand, water and gravel were performed. In the tests the same pools as in the unignited experiments were used and mainly ignition point in time and height were varied. The ignition settings were chosen on the basis of the results of the LH2 filling level measurements and the H2-concentration measurements of the unignited tests performed before.

The ignition was initiated during the evaporation phase after the second filling of the pool, since after the first filling vigorous boiling behaviour above the still rather warm substrate occurs. After the second filling the boiling behaviour above the precooled substrate is calmer and thus the filling level and concentration measurements give more precise values.

In the ignited experiments different degrees of damage were observed for the different substrate materials. For the substrates with a rather low porosity almost no or only minor damage was observed (see left part of Figure 5), while the experiment Gravel04 with highly porous substrate of gravel showed a complete destruction of the facility (see right image of Figure 5).

The reason for this exceptional behaviour, shown with a sequence of frames from the high speed video in Figure 6, is potentially condensed or frozen oxygen located in the free space between the stones of the substrate layer. This mixed with the refilled LH2 and finally reacted in the observed explosion.

Figure 5: Different degrees of damage to the facility observed in the ignited pool experiments for the different substrates

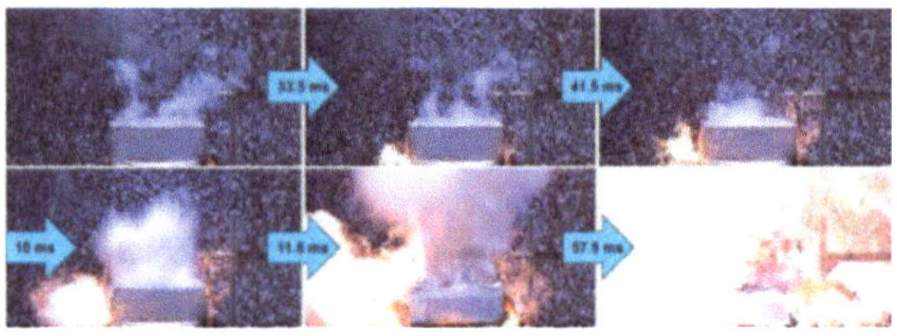

Figure 6: High speed video sequence of the experiment with ignition Gravel04 (PHOTRON High-Speed Camera, 2000 fps)

In summary, the observed evaporation rates provided valuable validation data for existing correlations and the observed massive explosion of the gravel substrate will initiate some modified recommendations for suitable ground materials for LH2 installations.

Transient cryogenic jet ignition

Transient cryogenic hydrogen jet fire behaviour, including scaling and radiation properties has been investigated with the small scale discharge facility DISCHA. The discharged hydrogen inventory varied with the initial pressure and temperature from 4.4 to 138 g. Similar as for the unignited discharge experiments done in 2019 [1] the release nozzle inner diameter was varied from 1-4 mm. Measurements consisted of background imaging system (BOS) combined with a high speed camera, fast pressure sensors and thermocouples. Additionally, a fast thermos-vision FLIR camera allowed monitoring the transient temperature fields. With mass flow rate and hydrogen distribution profiles known from the precursor unignited experiments, transient ignition phenomena and further flame development have been investigated with respect to maximum combustion pressure, temperature and heat flux radiation for model validation and hazard distance evaluation.

Unburned cold hydrogen jet was ignited with different delay time after jet initiation at different distances from the nozzle. Then, either a steady flame or a strong explosion with formation of a spherical shock wave might occur just after ignition (Figure 7). The over-pressure from 0.04 to 0.115 MPa corresponds to a visible shock wave velocity from 390 to 480 m/s measured by high speed BOS imaging.

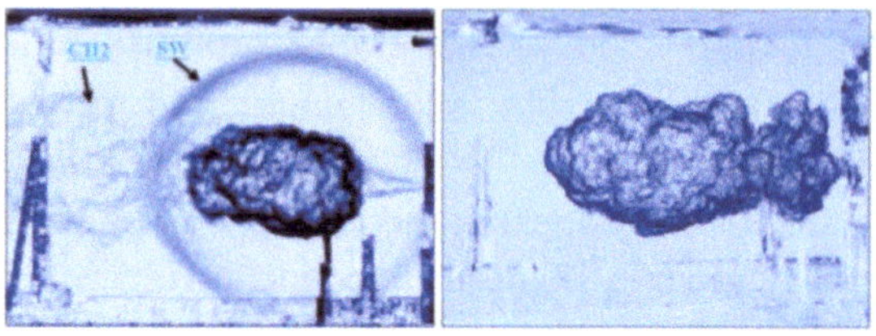

Figure 7: A shock wave formation (left) and a stationary jet fire (right) established under ignition of 4-mm nozzle and 20 MPa pressure hydrogen release: SW –shock wave; CH2 –unignited hydrogen

A sequence of frames with temperature distribution is obtained by thermo-vision FLIR-Camera (15 fps) (Figure 8). For ambient conditions 285K the local maximum combustion temperature changes from 1100 to 540K corresponding to maximum heat flux of 85 kW/m2. At the same time, the average integral heat flux of whole surface is about 6.5 kW/m2. At cryogenic temperature of 80K, the maximum temperature changes from 1330 to 710K corresponding to maximum heat flux of 177 kW/m2 in the center of jet fire. The average integral heat flux of whole surface is about 11 kW/m2. The reason of such difference is four times larger hydrogen inventory and also 2.5 times higher mass flow rate at cryogenic temperature leading to 1.3 times higher temperature, 2 times higher heat flux of flame radiation, 1.5 times larger flame length and 1.4 times longer release time.

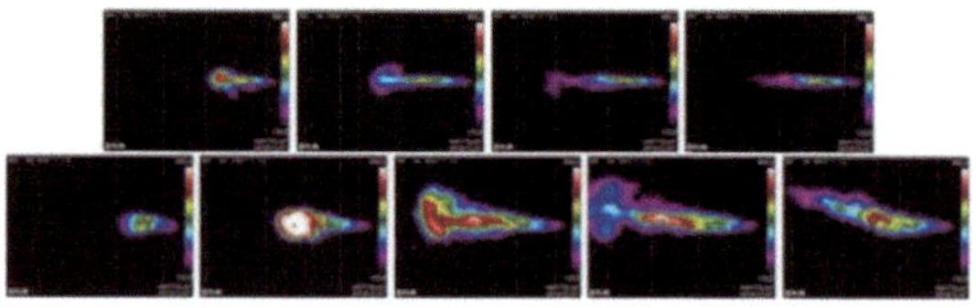

Figure 8: Temperature distribution within jet fire structure. Jet fire radiation for 2-mm nozzle and 10 MPa of initial pressure at T=285 K (upper sequence of frames) and T=80 K (lower sequence of frames).

Using the pressure records and the results of extensive optical observation with up to five cameras the ignition and combustion behaviour of the jets could be analysed. After ignition the flame either burns back to nozzle or not. This difference is important since in case the flame burns back to the nozzle it will continue burning, while in case the flame does not burn back to the nozzle it will quench after the ignition source is turned off. Using sensor records and optical observation on flame behaviour tests can be divided in two groups, where the flame burns back to the nozzle or not. When this data is combined with the results of the H2-concentration measurements of the unignited tests a clear trend can be observed. For very low H2-concentrations in the ignition position no ignition occurs, while for H2-concentrations lower than approx. 10 Vol% the jet is ignited but does not burn back to the nozzle. For H2-concentrations higher than approx. 10 Vol% in the ignition position the flame burns back to the nozzle.

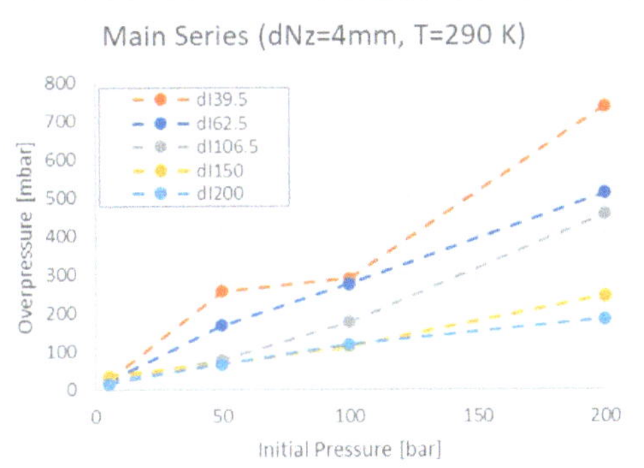

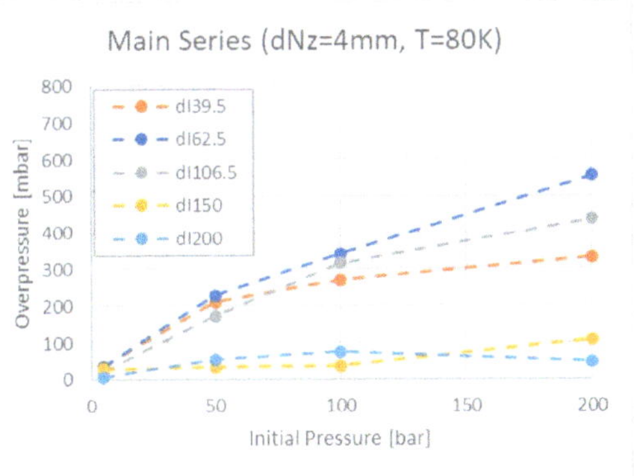

Figure 9. Maximum measured overpressures in experiments at ambient (top) and cryogenic temperature (bottom) of the discharged gases.

Concerning the combustion overpressures the following observations are made. In general, the maximum measured overpressures increase with increasing nozzle diameter, while the gas temperature in the reservoir seems to have no significant influence on the maximum pressure load. At ambient gas temperature the measured overpressures increase with decreasing ignition distance (Figure 9 top), while in the cryogenic tests the highest overpressures are measured for an ignition distance of 62.5 cm (Figure 9 bottom).

CryoTube Experiments

More than 100 experiments were made with the cryogenic combustion tube CRYOTUBE with L=5 m and 54-mm id and three blockage ratios BR = 0, 30 and 60% (Figure 10). About half of the experiments were made with hydrogen-air mixtures at cryogenic temperatures (approx. 80 to 130 K). The temperature was supported by the location of the tube under a layer of LN2 at T=77K (Figure 11). Subsonic, supersonic deflagrations and detonations were monitored for cryogenic hydrogen combustion by light sensors and pressure gauges located along the tube.

Figure 10: CRYOTUBE experiments mounted on a rack for the ambient warm pre-cursor tests.

Figure 11: CRYOTUBE immersed in a bath of LN2 with supporting structure.

The critical conditions for flame acceleration to the speed of sound or detonation were evaluated as a function of initial temperature. Particularly at 100 K, it shows a much higher hydrogen concentration of 16% H2 leading to sonic deflagration than that of 9.6%H2 predicted by advanced extrapolation of existing high temperature data before the tests. The correlation for the critical expansion ratio based on current experiments is quite simple and useful in a very wide range of initial temperatures (T = 90 - 650 K):

$$\sigma^* = 2200 \cdot T^{-1.12}$$

The run-up distance to the speed of sound or detonation in a smooth channel (BR=0) at cryogenic temperatures is two times shorter than at ambient temperature. For the first time, a steady-state flame propagation regime in a smooth channel with the speed of sound in combustion products was frequently observed in cases with suppressed detonation.

The detonation cell sizes at cryogenic temperature T = 100K have been evaluated on the basis of existing criteria for detonation onset in smoot and obstructed tubes and can be approximated by a polynomial function of hydrogen concentration [H2]:

$$\lambda[\mathrm{mm}] = 0.0006724[H2]^4 - 0.1039[H2]^3 + 6.0786[H2]^2 - 159.74[H2] + 1603.3$$

It appears that the detonation cell sizes for hydrogen-air mixtures at cryogenic temperature T=100K only insignificantly differ from that at ambient conditions. With the correct detonation cell size, all known DDT criteria can be used to assess the detonability of hydrogen – air mixtures in different geometries and scales at cryogenic temperatures.

The maximum combustion pressure at cryogenic temperatures is 2-3 times higher than that for ambient conditions. Theoretically, the maximum combustion pressure for sonic deflagration, which is roughly equal to adiabatic combustion pressure, is proportional to the temperature ratio T0/T, where T0 =293K and T is the cryogenic temperature. Similarly, the maximum detonation pressure can also be predicted. It demonstrates a very high level of hazard under cryogenic hydrogen combustion. Even a sonic deflagration can be 1.5 times more dangerous than the detonation of the same hydrogen-air mixture at ambient conditions.

A broader overview on all PRESLHY results may be found in [2], [3] and in particular in [4].

Partner

We would thank Andreas Friedrich, Joachim Grune, Anke Veser from Pro-Science for the good work, which they have done.

References

[1] Jordan, T. PRESLHY Experiment series E3.1 (Cryogenic Hydrogen Blow-down/Discharge) results – part A "high pressure", https://publikationen.bibliothek.kit.edu/1000096833

[2] Jordan, T., Safe Use of LH2 in Non-industrial Settings, Contributions of the FCH JU project PRESLHY 13th plenary SFEM Working Group HYDROGEN, 28 January 2020, Brussels, Belgium (https://hysafe.info/wp-content/uploads/sites/3/2020/02/20200128_SFEMWorkingGroupHYDROGEN_30min.pdf)

[3] Jordan, T.; Hecht, E.; Odegard, A.; Low T LH2/cryogenic related progress in Safety Research HySafe Research Priorities Workshop, 30 October 2020, online presentation (https://hysafe.info/wp-content/uploads/sites/3/2020/11/ResearchPriorities_LH2_Jordan_Hecht.pdf)

[4] https://preslhy.eu/meetings/dissemination-conference/

First experimental results of a supercritical Organic Rankine cycle using propane as working fluid

Luciano Gardella, Hans-Joachim Wiemer

In geothermal power generation, hot water is pumped at around 100-200°C from deep layers of rock in the earth's interior and converted into electrical power via a cycle process, which is usually fed into the public power grid. Electricity generation for these low-temperature cycles is nowadays usually realized in low to med enthalpy wells via an ORC process (Organic Rankine Cycle).

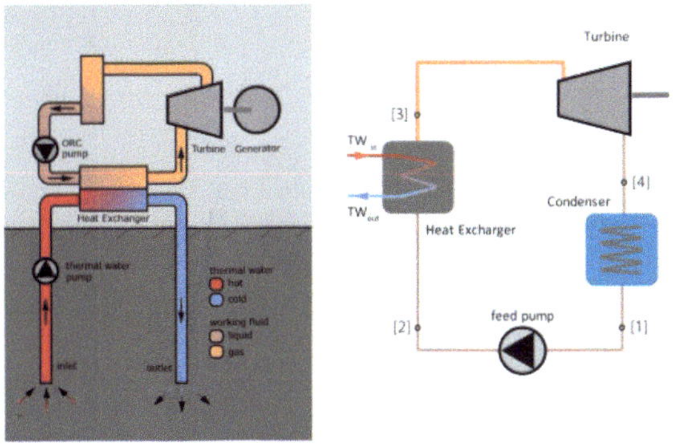

Fig. 1: ORC process

The variation of the boundary conditions and the working fluids used has a large influence on the net electricity yield and thus on the economic efficiency of a geothermal power plant.

The approach followed by the ITES-EVT group is to maximize electricity production by selecting a heat transfer fluid suitable for the site-specific boundary conditions in Europe and a supercritical live steam pressure. In previous studies [1],[3] Vetter et al. have shown that significant performance improvements are possible at supercritical conditions.

The working fluid selection has taken into account environmental and operational criteria, such as high thermal conductibility, low specific volume, high chemical stability, low corrosiveness, low flammability, toxicity, low Ozone Depletion Potential (ODP) and Global Warming Potential (GWP) [2].

In order to optimize and validate this simulation model we built a supercritical ORC (MoNiKa).

The MoNiKa power plant

MoNiKa (Modular low-temperature cycle Karlsruhe) is a facility built at KIT campus north with the idea of studying and optimizing the ORC process. This installation is a small and compact power plant. It was designed as a modular installation to allow the study and investigation of different components and operational parameter for the research of geothermal power generation from low-temperature heat sources.

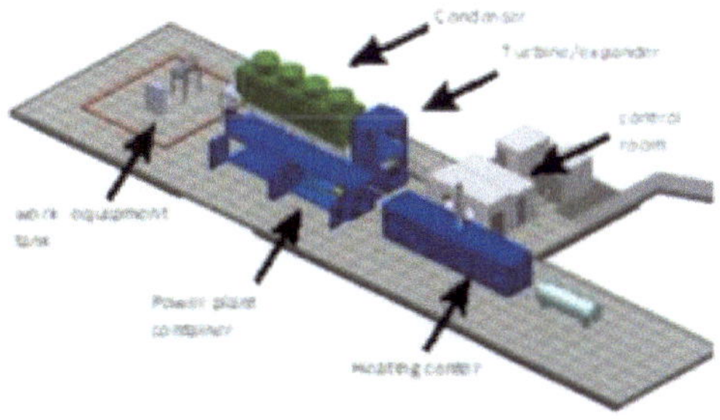

Fig. 2: MoNiKa -ORC power plant

Components description

The pumping system is compound by two pumps. The main pump is a LEWA triplex M514US G3G. It is a piston pump of max. 75 kW with a maximum mass flow of 3.6 kg/s and a design pressure of 6.5 MPa (in this will be referenced as main pump).

The heat exchanger manufactured by Gesmex, is the connection between both cycles, (thermal water and organic). It is a counter-cross-flow heat exchanger designed to work in a supercritical regime. This component consists of 200 circular welded stainless steel plates grouped in five stage. The design thermal power installed is 1000 kW for full load operation

The condenser manufactured by KÜHLTURMKARLSRUHE has a support sub frame with 3.5 m of height. It is located between the exit of the turbine (or throttling valve), and the propane tank. However, the condenser is prepared to work with water spray. The heat exchange areas are built symmetrically in "V" configuration. They include three chambers. Each one is equipped with a vertical fan (impeller and diffusor) of 2.5m diameter. The power consumption of each motor is max.13 kW at 322 RPM, and a maximum volume flow rate of 44 m^3/s each and in total 39 kW - 132 m^3/s.

The last component of the circuit is the turbine from M+M. The axial turbine has four stages with a nominal rotational speed of 9996 1/min. The 150 MVA synchronous generator is connected to the turbine with a gearbox. The design requirements assumed an isentropic efficiency at full load of 0.8 and a vapour quality limit at the turbine outlet of 0.9.

Fig. 3:Propane tank und Condenser

Fig. 4:Main pump and Plate heat exchanger

Fig. 5: Turbine with gearbox and generator

In addition to these main components, the following auxiliary units are also important for the power plant operation.

Support pump from Grundfos, which ensures the upstream pressure to the main pump.

Propane separator from Borsig for separating the nitrogen propane mixture from the gas charged in the turbine shaft seal. The separated propane is returned to the propane tank.

Propane tank with a volume of 2.4 m³, also an additional storage tank for refilling propane.

Furthermore, we have a 7-m³ fuel oil tank to supply the heating station.

The control and regulation technology container with a redundant S7-416 PLC and the SIEMENS power plant software T3000.

The Control room container with three workstations for power plant and instrumentation monitoring.

Scientific measuring points for the acquisition of additional parameters such as ambient temperature and humidity, pressure difference condenser, temperature and velocity field condenser as well as an optical access to the propane circuit for flow characterization.

Instrumentation

The sensor technology installed in MoNiKa fulfills two requirements: the first for power plant control and regulation, the second as a platform for scientific studies conducted in the plant. Therefore, the plant is much more extensively instrumented than in a normal power plant. For balancing purposes, sensors for temperature and pressure are installed along the entire cycle, i.e., at the inlet and outlet of each component.

The mass flow rate and density of the working fluid are measured between the outlet of the main pump and the inlet of the heat exchanger using an E+H Promass 83F sensor. This sensor uses Coriolis forces and resonant frequency to provide a direct measurement of mass flow, velocity and density.

A WIKA TR34 Class A-PT100 is used to measure temperatures at all points in the system. These sensors are connected as 4-wire resistance thermometers and have an operating range of -50 to 250 °C. These thermometers are characterized by a compact design, are resistant to vibration and have a fast time response for operating measuring points. Type K thermocouples are used in areas with increased time requirements.

For pressure measurements at the MoNiKa plant we use the Vegabar 81 and 82. These pressure transmitters are universally applicable for the measurement of gases, vapors and liquids. They have a ceramic measuring cell, which allows the sensor to behave well in corrosive and hot environments. The main difference between purchased sensors is the temperature range in which they can operate. In addition, the pressure range in which they operate is different within the ORC cycle, resulting in pressure measurements in the low-pressure section having higher absolute accuracy than in the high-pressure section.

The sensors are in direct contact with the liquid (propane and water). There is no sleeve or cap between them. This direct contact allows better measurement of the fluid properties, but makes it impossible to remove the sensor from the pipeline for maintenance without first draining it. For this purpose, MoNiKa's measurement system is designed with redundancy, which means that two or three sensors of the same type are installed at each measurement point. This redundant configuration makes it possible to increase the availability of the measurement technology, which on the one hand increases safety during operation and on the other hand helps to detect systematic errors.

To measure the environmental conditions, the system is equipped with an EE33 series humidity and temperature transmitter. In addition, a set of type k thermocouples is installed in each chamber of the condenser.

Tab. 1: List of Sensors

Magnitude	Model	Measurement Range	Max Absolute error
Pressure	Vegabar81	0 to 10 MPa	0,02 MPa
Pressure	Vegabar82	0 to 10 MPa	0,01 Mpa
Pressure	Vegabar82	-1 to 10 MPa	0,01 Mpa
Pressure	Vegabar81	-1 to 2,5 MPa	0,005 MPa
Pressure	Vegabar82	-1 to 2,5 MPa	0,003 Mpa
mass flow	Promass 83F	-	0,005 kg/s
density	Promass 83F	-	0,01 kg/L
Temperature	TR34 class A	-50 to 250 °C	0,4 °C

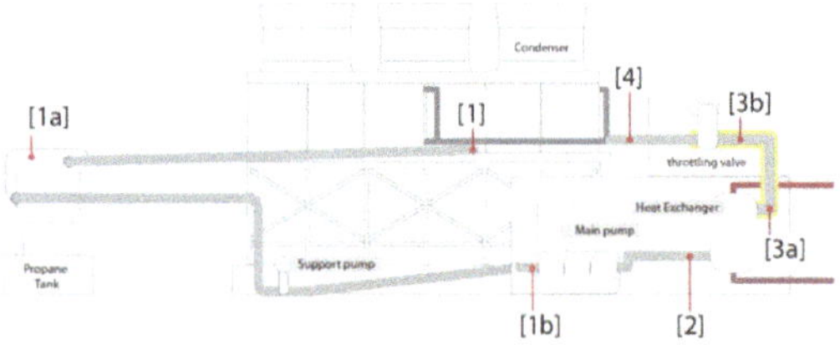

Fig. 6: Operational measurement points at MoNiKa

The accuracy of the data acquisition system is mainly affected by the analog to digital converter with a resolution of 14 bits, which corresponds to an electrical current value of ±1.221 µA.

MoNiKa's PROFIBUS has a cycle time of 100ms. The usual average data sampling time was 1 sample/ Sec. For long test runs we use 1 sample / 10 sec. because the smaller value exceed the maximal file size of the standard data export function in T3000.

Operational design parameter

The plant is a binary cycle, where the geothermal heat source is simulated. A hot water boiler with one MW of thermal power heats the water at the site, simulating the geothermal water. We are able to modify the temperature and mass flow rate at the water loop in order to have a range of input conditions and provide different scenarios. The following table shows the expected design parameter at full load operation.

Tab.2: Designed operation values.

ORC Cycle	
Propane as Organic working fluid	
Live steam point [3]	5.5 MPa and 117 °C.
ORC mass flow	2.9 kg/s
Turbine isentropic efficiency	80%
Fluid quality (at the condenser inlet)	95%
Pumps isentropic efficiency	70%
Thermal Water Cycle	
mass flow	2.4 kg/s
Temperature in	150 °C
Temperature out	47 °C
Pressure in	0.9 MPa
Power	
Thermal power:	~ 1000 kW
Heat released to the ambient	~ 930 kW
Gross Power generation	~ 150 kW
Net Power generation	~ 110 kW
Thermal efficiency	~ 15 %

Simulation Software GESI

GESI (Geothermal Simulation) Vers.2.3.6b [5] is an in-house program that has been developed in MATLAB by the ITES (Institute for Thermal Energy Technology and Safety) for studying and optimizing the thermodynamic process of the Organic Rankine Cycle (ORC). The software use the physical property data taken from REFPROP 9.0, from the National Institute of Standards and Technology [6].

This tool solve the energy and mass balance of the ORC Power Plant in stationary regime. As an input the definition of the thermal water values, selection of the operational points, the ambient characteristic and the equipment is necessary. The tool provides as result the major values of the whole process and deliver diagrams e.g. temperature- and entropy diagram.

The results of the simulation allows to validate the power plant model and to determine the efficiency of the pump and turbine in full- and part load operation.

Test run

MoNiKa is a research platform that was completed at the end of 2018. The first tests were performed in November 2019, without the turbine, operating in bypass. The focus of these tests was to investigate the entire cycle and all the components involved.

Despite the fact that the plant was fully completed, all components are installed and functional, some control loops of the control system were not fully optimized and automated at this time. This means that most of the test runs have been manually controlled.

With the turbine tests in January 2021, the entire commissioning of the plant was completed.

The optimization of the control and regulation circuits is a task in parallel to the actual test operation of the plant and will be further developed on an ongoing basis.

The first longer bypass test took place on 23.01.2020 and lasted 20 hours with an ORC mass flow of 100%,70% and 50 %. The table 3 shows the results for the 100% case.

Table 3: test run results 100% load

	Full Load 100%$\dot{m}$	**Accuracy**	**Unit**
ORC cycle			
m_{ORC}	2.87	±0.03	kg/s
$T_{AIR\ cond\ out}$	6.79	±0.75	°C
T[1]	13.8	±0.2	°C
p[1]	0.722	±0.003	MPa
T[1a]	14.3	±0.2	°C
p[1a]	0.721	±0.003	MPa
T[1b]	14.0	±0.2	°C
p[1b]	0.777	±0.003	MPa
T[2]	17.0	±0.2	°C
p[2]	5.53	±0.01	MPa
T[3a]	108.1	±0.4	°C
p[3a]	5.52	±0.02	MPa
T[3b]	108.3	±0.4	°C
p[3b]	5.50	±0.02	MPa
T[4]	14.6	±0.2	°C
p[4]	0.72	±0.01	MPa
Water Cycle			
m_{tw}	2.4	±0.02	kg/s
ρ_{tw}	0.9	±0.01	kg/L
$T_{tw\ in}$	150.2	±0.5	°C
$T_{tw\ out}$	61.74	±0.3	
Ambient			
T_{amb}	0.98	±0.75	°C
Main pump			
P_{hyd}	26	±2	kW
P_{Elec}	27.5	±0.2	kW
Support pump			
P_{hyd}	0.31	±0.006	kW
P_{Elec}	1.06	±0.02	kW
Heat transfer			
He Ex	894	±8	kW

With these measured data we can calculate the enthalpy and entropy values for the four cycle points as mentioned in Fig.1.

Table 4: Test run enthalpy and entropy results with the estimation for point 4.

ORC points	Enthalpy	Entropy	Quality
	kJ/kg	kJ/kg K	-
[1]	235	1.12	Subcooled
[2]	245	1.13	Subcooled
[3]	557	2.04	Super-heated
[4]	-	-	-
[4]* estimated	557	2.24	0.925

At point 4 the bypass outlet/condenser inlet we have only pressure and temperature sensors. In the two phase region we need an additional parameter to define the enthalpy or entropy values.

So we can estimate the enthalpy with the assumption of an isenthalpic throttling from point 3 to 4 (bypass valve) and neglect the heat loss.

The heat transfer released from the water cycle is determined to 899 kW. The amount of heat absorbed by the propane is 886 kW with a deviation of 10 kW. This mean that we have a heat transfer efficiency of 98% ±1.5%.

Nevertheless, the total heat transfer is about 10% less than the design value. This circumstance significantly affects the performance of the entire power plant. It is not only the power generation that is affected. The biggest impact is the quality of the fluid after expansion.

The pressure loss of the heat exchanger is less than the accuracy of the pressure sensors.

The designed minimal temperature difference of the heat exchanger was 10 K and we measure a value of 33.5 K.

The total pump efficiency (hydraulic power / electric power) for the main pump was above 90% in the range of 50% -100% of ORC mass flow with a maximum value of 97%.

For the support pump, we determine the total efficiency to 29% -33%.

In order to balance the released heat in the condenser, the heat release on the airside was verified.

Using the air temperature measurement and under the boundary condition that the fans were operating at full power, the heat released was calculated and then the thermodynamic values of the state point 4.

The result is inconsistent, since from the simulations a heat release of 1100 kW was expected and in the test run we measured only about 930 kW as maximum released power.

These results show a similar steam quality after throttling and a slight sub cooling of the propane at the condenser outlet, which is not considered in the simulation. However, the difference in enthalpy and entropy is only 0.1% and does not explain the lower heat release.

Additional investigations have shown that the propane tank, due to the large propane mass, determines the condensation pressure at the condenser outlet and thus also the minimum temperature.

More detailed investigations of the condenser will be carried out in the future.

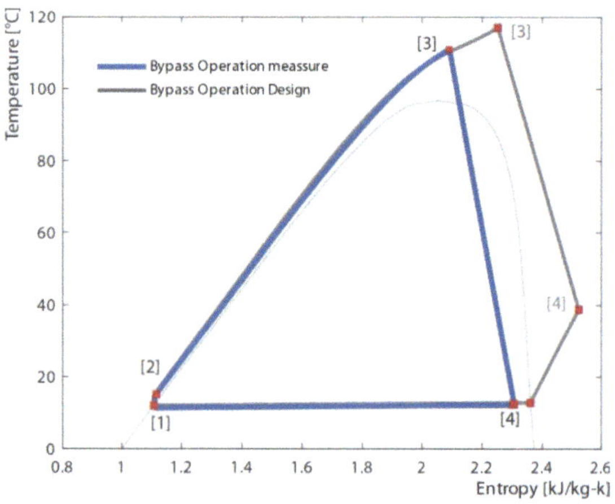

Fig. 7: Ts-Diagram of the ORC in bypass operation in comparison to the designed bypass operation.

During turbine commissioning in December 2020, significant vibration amplitudes occurred in the turbine and gearbox bearings, exceed the predefined limits and causing the turbine to trip.

Only after readjustment and replacement of the vibration sensors, it was possible to run the turbine up to nominal speed.

Unfortunately, the synchronization of the generator only worked after a manual adjustment of the excitation current controller, so that we were able to feed electricity into the grid for the first time on January 25, 2021.

In this test of commissioning, the turbine was loaded up to an electrical gross power output of 88.6 kW.

Summary

The MoNiKa cycle in stationary bypass operation was studied. As a first step, we study the installed sensors in the facility in order to determine the accuracy of the measurements.

Experiments were carried out at three operating points 50%, 70% and 100% of ORC mass flow and analyzing the behavior of the entire power plant as well as the individual components.

The 10% less heat transfer at the heat exchanger in comparison to the design values makes it necessary to redefine the optimal full load point and limit the range of operation due to the minimal allowable vapor quality at the turbine in- and outlet.

The adaption of ORC turbine inlet live steam conditions with an ORC mass flow of 2.7 kg/s, which is only 93% of the designed mass flow, is the result of the simulations performed with GESI with the corrections already discussed, and considering the limits of the thermal water cycle.

We completed the commissioning of the Monika plant with the successful power grid supply. In the further experimental operation, the focus will be on investigations and optimization of the single components and the full system performance.

References

[1] Vetter, Christian; Wiemer, Hans-Joachim and Kuhn, Dietmar; Comparison of sub- and supercritical Organic Rankine Cycles for power generation from low-temperature/low-enthalpy geothermal wells, considering specific net power output and efficiency. Applied Thermal Engineering 51 (2013) 1-2, S. 871–879

[2] DiPippo, R.: Geothermal power plants. Principles, applications, case studies and environmental impact / Ronald DiPippo. Amsterdam: Butterworth-Heinemann 2015

[3] Vetter, Christian; Wiemer, H. Joachim; Dynamic Simulation of a Supercritical ORC using Low-Temperature Geothermal Heat. Melbourne, Australia (2015)

[4] Vetter, Christian; Parameterstudie zur Simulation von Niedertemperatur-Kreisprozessen (KIT scientific reports) 2011, ISBN 978-3866446731

[5] ITES Institute for Thermal Energy Technology and Safety: GESI. Geothermal Simulation. 2018

[6] REFPROP. Reference Fluid Thermodynamic and Transport Properties Database. NIST National Institute of Standards and Technology (2018)

Group: Framatome Professional School

Annual report of the Framatome Professional School

Andreas Class

The Framatome Professional School (FPS) organizes training courses in safety technology, thermal hydraulics, reactor physics, stress analysis, and hydrogen technology etc. Courses are tailored for engineers specializing in new topics within Framatome and they are open for students from KIT and other universities. Moreover, through the regular exchange between KIT and Framatome public funding opportunities for joint proposals are identified.

The FPS coordinates the InnoEnergy Master Schools Programmme "Energy Technologies" which offers a broad-based education in a variety of key engineering disciplines. A focus lays on the inter-disciplinary aspects of the sustainable energy sector and an in-depth understanding of the role of innovation and entrepreneurship in the future energy industry. Students receive a Master's degree from two of the four participating universities (IST: Instituto Superior Técnico in Portugal, Grenoble INP: Institute of Technology in France, UU: Uppsala University in Sweden, or KIT). The FPS also organizes ex-curricular activities for students such as the annual SiemensEnergy New Energy Challenge.

The FPS has a long history of applied research and strong links to other ITES-groups exist. The FPS applies Computational Fluid Dynamics and System Codes, develops Reduced Order Models for Uncertainty Quantification and mesh-free modelling of three-dimensional thermohydraulic components, and exploits methods of applied mathematics such as asymptotic series expansions to complex physical problems.

The applied focus of the research of the FPS led to inventions in 2020. A lab-on-a-chip STOKES II developed by P. Marthaler and A. Class allows driving μm-scale fluid flow on a chip in a freely programmable fashion. This provides a new platform technology for complex diagnostics and chemical synthesis tasks.

The FPS invention BattMarines addresses electric energy storage in a power-heat-power concept.

In the following sections a number of activities of the FPS are summarized.

CFD Validation of heated rod bundles in heavy liquid metals

Abdalla Batta, Karsten Litfin, Andreas Class

In our group, validation of CFD models for thermal-hydraulic simulation is an essential part of our work. Recently many studies are performed for supporting the design of next generation reactors like the "Multi-purpose hybrid research reactor for high-tech applications (MYRRHA)" planned at SCK-CEN, Belgium. For this purpose a series of EU-projects have been started in the last years like EUROTRANS (2005-2010), THINS (2010-2015), SEARCH (2011-2015) and MAXSIMA (2012-2018), to name a few.

The ongoing Project PATRICIA (2020-2024) aims to support safety studies for heavy liquid metal cooled fast reactor systems. The effect of a well-defined porous blockage in a wire spaced 19-pin rod bundle and the effect of rod bundle deformation will be experimentally investigated at KALLA with numerical support from our group. Further numerical support will given for the project PASCAL (2020-2024)

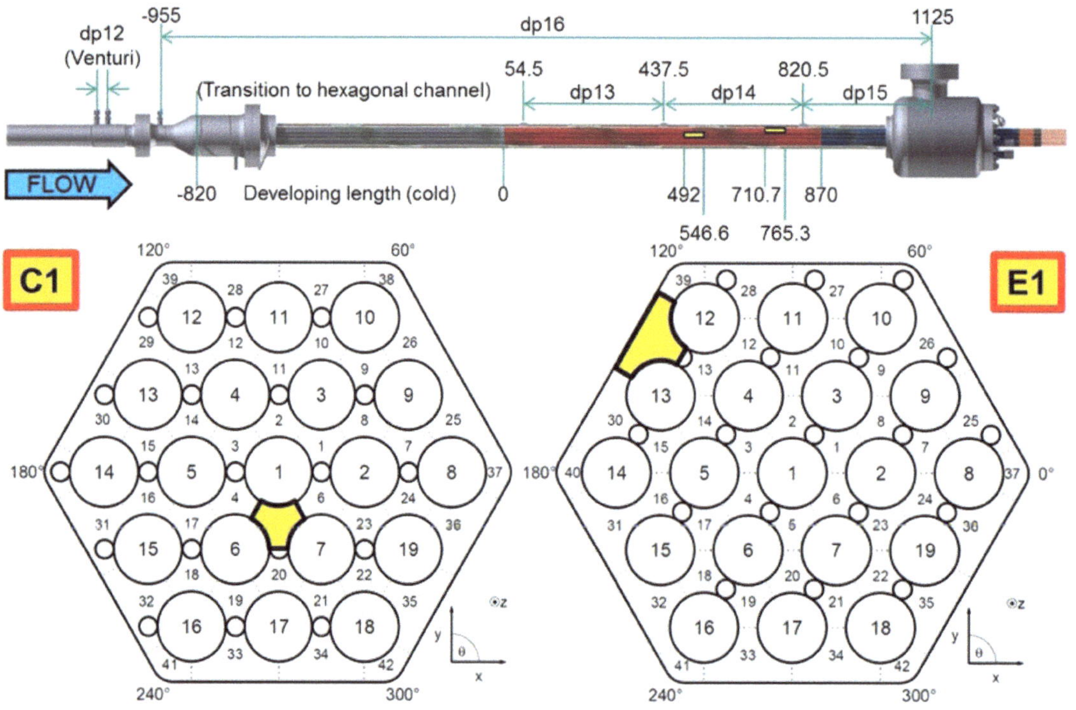

Figure 1: Schematic side view of a rod bundle test section in heavy liquid metal with two small blockage elements. All axial positions are expressed in mm, referring to the onset of the heated zone (z=0).

where the effect of deformation on the local flow and temperature field will be studied.

An example for the validation of experimental results obtained within the EU FP7 project "Methodology, Analysis and eXperiments for the Safety In MYRRHA Assessment (MAXSIMA) project will be presented in the following subsection. More details about the project can be found elsewhere [1], [2], [3]. Figure 1 shows the test section with two blockage elements (marked in yellow) installed at selected locations in the heated section (marked in red). Details of the study are given in [4].

Model validation

In this example, we present a comparison of the experimental results to CFD calculations. To quantify the impact of the blockage (I) onto the local temperature we define the impact factor as:

$$I = 100\frac{T_{blocked} - T_{unblocked}}{T_{blocked} - T_{inlet}}$$

The impact factor is presented for both the experimental and numerical data in Table 1. The evaluation employs temperature values measured behind the first blockage (Central one). From experimental data, one can see that wall temperature is influenced with maximum 10.9 % at the thermocouple $T_{ML2\_R16\_240}$, where the blockage impact on fluid reaches 22.7 % at T_{ML2_S41}. Considering the numerical results, it can be seen that the blockage impact is smaller than the experimental values and take place at different location. This shows that current models are not yet adequate to predict flow mixing around blockages.

For the comparison between numerical and experimental local results, the error (Err) can be calculated by:

$$Err = 100\frac{T_{experimental} - T_{numerical}}{T_{experimental} - T_{inlet}}$$

Table 1: Experimental and numerical temperature data for case 157 and 24 of the blocked rod bundle experiment at KALLA. All temperatures in [°C]

	Experimental			Numerical			Num. error comp. to Exp.	
type	temp. blocked	temp. Unblocked	I [%]	temp. blocked	temp. unblocked	I [%]	Err %	Err %
Case number	157	24		157	24		157	24
T_ML2_R1_270	278.31	278.64	-0.4	285.7	283.6	2.5	9.5	6.3
T_ML2_R6_330	276.64	276.56	0.1	281.6	281.0	0.7	6.4	5.8
T_ML2_R7_210	269.04	269.67	-0.9	274.7	278.7	-5.0	8.2	12.9
T_ML2_R16_240	257.11	251.51	10.9	246.9	243.1	8.7	-17.9	-16.2
T_ML2_R16_330	266.81	267.83	-1.5	264.1	268.5	-6.4	-4.1	0.9
T_ML2_R17_30	264.27	266.41	-3.2	269.4	271.7	-3.3	8.0	8.0
T_ML2_R18_150	263.95	266.2	-3.4	276.4	274.3	2.8	19.4	12.2
T_ML2_S5	262.83	260.75	3.4	274.4	266.3	12.3	18.5	9.1
T_ML2_S20	270.01	259.82	17.0	262.0	262.7	-1.1	-11.4	4.8
T_ML2_S21	261.59	257.19	7.7	262.0	258.9	5.2	0.7	3.0
T_ML2_S33	245.64	246.52	-1.9	259.3	249.9	18.9	30.0	7.3
T_ML2_S41	237.57	230.61	22.7	225.8	223.3	10.7	-31.2	-23.8

The calculated error values are given in the last two columns of the Table. Note that the maximum error is observed for the fluid temperatures measured near the outer wall. This can - to some extent – be related to heat losses at the outer wall of the bundle which has been neglected in the simulation. However, the dominant reason is a weak capability of the model to predict turbulent liquid-metal flow with obstacles and strong circulation.

Based on the presented data the Nusselt number can be calculate yielding for the experimental data: $Nu_{blocked} = 10.54$ and $Nu_{unblocked} = 11.03$. The value for $Nu_{unblocked}$ is similar to the value $Nu = 11.2$ calculated based on comprehensive data. From the numerical data $Nu_{blocked} = 9.73$ and $Nu_{unblocked} = 9.88$ are obtained. Comparing experimental to numerical results for Nusselt numbers better agreement than for local temperatures is found. Thus, the blockage has a minor effect on the calculated Nusselt number, since the latter is based on the average temperature. Note that the predicted temperature error scatters exhibiting negative and positive values. Therefore, in the averaged values the difference relative to simulated results is less when comparing local data to experimental data. Detailed discussion on the experimental results can be found in [5]

References

[1] Batta, A.; Class, A.; Pacio, J.; Numerical analysis of a LBE-cooled blocked 19-pin hexagonal wire wrapped rod bundle experiment carried out at KIT-KALLA within EC-FP7 project MAXSIMA" Paper ID 20532, Nureth 17, 2017.

[2] Pacio, J.; Daubner, M.; Fellmoser, F.; Litfin, K.; Wetzel, T.; Experimental study of heavy-liquid metal (LBE) flow and heat transfer along a hexagonal 19-rod bundle with wire spacers, Nuclear Engineering and Design, 301, pp. 111 – 127 (2016).

[3] Pacio, J., et.al.; Final report on partially blocked wire wrapped LBE rod bundle experiment, Technical Report, MAXSIMA Deliverable D3.4,2017.

[4] Batta, A; Class, A.; Validation for CFD thermalhydraulic simulation for liquid metal cooled blocked 19-pin hexagonal wire wrapped rod bundle experiment carried out at KIT-KALLA , AMNT19 , 2019.

[5] Pacio, J.; Daubner, M.; Fellmoser, F.; Litfin, K. und Wetzel, T.; Heat transfer experiment in a partially (internally) blocked 19-rod bundle with wire spacers cooled by LBE Nuclear Engineering and Design, 2018, 330, 225 – 240.

Creation of a reduced order model for the k - ε model

Jorge Yanez, Andreas Class

Introduction

Nuclear technology is a sector in which safety is insured in a very strict manner. In these frames, is becomes crucial to analyze the statistical trustworthiness of Computational Fluid Dynamics (CFD) results. That is to assess the effect of the bias and the unavoidable random variation and aleatory disposition of the initial conditions of a realistic scenario. For this task Reduced Order Models (ROM) come to hand [3]. Here we address the creation of a ROM for the standard k-ϵ- turbulence model as it represents the "work horse" of CFD analysis. This is not a simple task since the k-ϵ model equations exhibit multiple non-linearities.

Theoretical model

We start with the formulation of the Navier-Stokes equation in its Boussinesq approximation and the complementary k-ϵ equations[4]. In our methodology [2], we utilize the Method of Snapshots [1] to create a simplified model. That is, we obtain the transient solution of our equations with the high fidelity solver STAR-CCM at times $\{t_1, \dots, t_n\}$. We form with them a snapshot matrix Y. We carry out its Singular Value Decomposition, $Y = V\Sigma G^T$. We project the n dimensional vectorial space in a–reduced–sub-space of dimension $n - j$ dropping the last j vectors in V. The first $n - j$ vectors of V create a vectorial space at most $\sqrt{\sum_{i=j}^{n} \| \sigma_i \|^2}$ distant form the original one.

To develop the reduced order model, we write the k-ϵ equations in terms of the reduced order basis. We follow the same rationale as [3]. After a quite complex algebra, the model adopts the form,

$$\left(\mathbb{D} + \frac{\Delta t}{2}\mathbb{E} + \frac{\Delta t}{2}\mathbb{F}\mathbb{J}\right)\begin{pmatrix} \overrightarrow{\delta u} \\ \overrightarrow{\delta T} \\ \overrightarrow{\delta k} \\ \overrightarrow{\delta \epsilon} \end{pmatrix} - \Delta t \mathbb{G} = 0$$

Note that $\mathbb{D}$, $\mathbb{E}$, $\mathbb{F}$, $\mathbb{G}$ and $\mathbb{J}$ are matrixes with a much smaller dimension compared to the degrees of freedom of the high fidelity calculation. They depend on the dimension of the reduced basis. Yet, the matrix entries depend on the comprehensive high-fidelity calculation. All matrix entries are calculated once during the post processing of the high-fidelity results, and subsequently stored.

Application

We apply our construct to a convection case of domain 10x10x1 meter and resolution 0.1 meter. We consider the volume completely isolated for the whole set of variables, with zero gradient perpendicular to the boundaries.

Considering the snapshots generated and a maximum distance of 10^{-7} to the projection space, the dimensions of the reduced space is 25 considering all variables of the system. Note that that means that the--full--matrices of the model are in fact of size 25×25 in total, compared with the dimension 100.000x100.000 of the—sparse--matrices of the original calculation for each variable.

The model has been implemented in an in house Python code. The results of the integration of the ROM compared with the original CFD results can be seen in Figure 2. The CFD result has been projected into the reduced basis. The amplitudes in this basis of the ROM and the CFD can be directly compared. We see an excellent agreement between CFD and ROM.

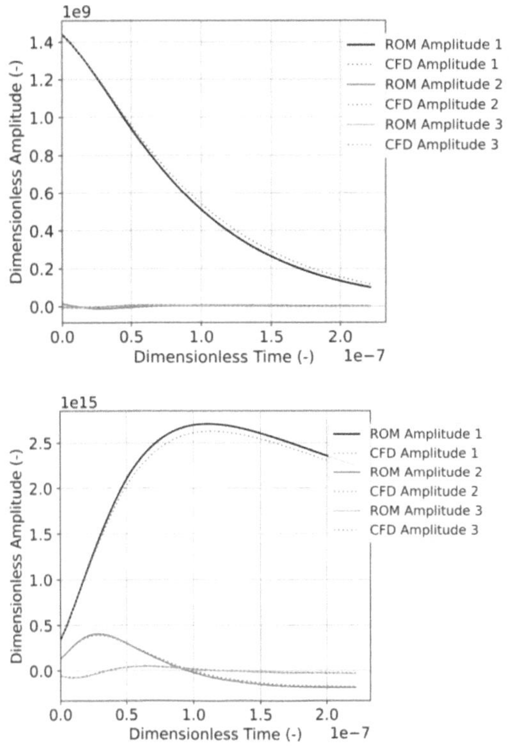

Figure 2 Comparing the reduced order model (ROM) with original CFD calculations show an excellent agreement. (top: velocity; bottom: k)

Conclusions

We have shortly summarized the procedure we have followed for the creation of a reduced order model of the k-ϵ turbulence model based on Proper Orthogonal Decomposition and Galerkin projection. The model finally adopts a simple expression.

In the whole derivation procedure, on-line operations, that is, those to be done during the integration of the reduced model, have the complexity of the reduced basis. Only off-line operations, done once during the post processing the results of CFD have the complexity of the original basis.

The construct obtained shows a very good performance reproducing the calculation proposed to illustrate its capabilities.

References

[1] Berkooz, Gal; Holmes, Philip and Lumley, John L; 1993. "The Proper Orthogonal Decomposition in the Analysis of Turbulent Flows." Annual Review of Fluid Mechanics 25 (1): 539–75.

[2] Yáñez, Escanciano Jorge and Class, Andreas G.; 2019. "POD-Galerkin Modeling of a Heated Pool." Progress in Nuclear Energy 113: 196–205.

[3] Quarteroni, Alfio; Gianluigi Rozza, et al.; 2014. Reduced Order Methods for Modeling and Computational Reduction. Vol. 9. Springer.

[4] "STAR Ccm+ Users Manual." n.d.

A free programmable lab-on-a-chip

Philipp Marthaler, Andreas Class

Background

Lab-on-a-chip systems are utilized in many fields for rapid and reproducible chemical analyses on a microscale. Especially in medical technology, such tools support diagnoses and research. The design of most systems, however, is customized for one specific application. Important application cases are point-of-care diagnostics or DNA sequencing. Such single-application systems are characterized by networks of micro-channels. Channel networks ranging from rudimentary (single-use systems) to complex (covering multiple use cases) are available.

In the described systems, probes are embedded and transported in a fluid environment. The carrier fluid which contains the diffuse samples is excited to flow by induced-charge electro-osmosis. Most common is the excitation with AC potentials applied via electrodes inside the channel wall. In contrast to the application of DC fields, that method has several advantages. In particular, it is applicable in more complex geometries and creates more continuous flow patterns.

Innovative approach

We developed a novel lab-on-a-chip system that does not depend on channel networks. The carrier fluid with diffuse probes is transported using AC electro-osmosis with traveling waves, also known as traveling wave electro-osmosis.

The central element of the development of that system is a spectral code that simulates the underlying electro-hydrodynamic phenomena. With that code, an earlier ITES patent [1] was examined for its effectiveness and feasibility. Based on the results, the excitation method was improved from alternating current to traveling waves. The code was further utilized to find the optimal operating parameters for the system based on fluid properties, ion species, and concentration. Compared to most available systems, high transport velocities were reached.

The optimized excitation method can now be applied in a redesigned 1mm x 1mm chip (schematically shown in Figure 3) with a free flooded surface. Channels are not necessary for this system. Thus, many processes can be performed on one chip. Those processes depend mainly on the software implementation and only to a minor part on the chip geometry.

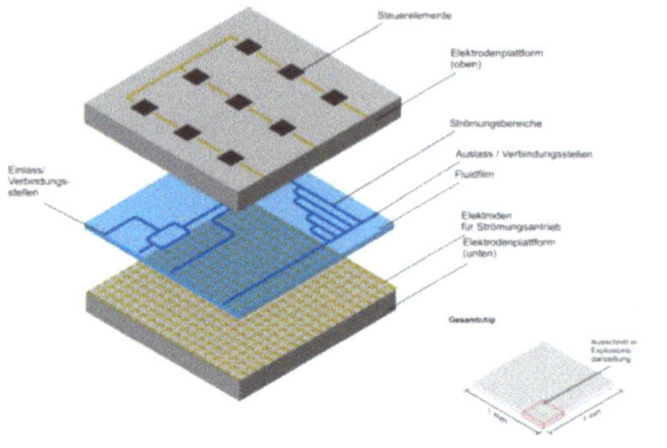

Figure 3 Schematic sketch of the chip structure. Dark blue paths represent example flow patterns in the flooded surface (light blue). The electrodes are controlled by single control units (black squares) and are responsible for flow excitation.

Besides flow excitation and probe transport, the design of a wire network inside the chip supplying a large number of electrodes was an immense challenge. For the prototype design and numerical verification, Philipp Marthaler and Andreas Class won the 2nd prize at the KIT innovation contest NEULAND [2]. The award ceremony is shown in Figure 4.

Figure 4 Award ceremony of the NEULAND contest.

References

[1] Class, Andreas; Barz, Dominik; (2006): Arrangement for generating liquid flows and/or particle flows, method for producing and operating said arrangement and use of the latter, (WO 2007/090531 A1), World Intellectual Property Organization

[2] Marthaler, Philipp; Simon, Marie; (2020): Interview with the nominees of the NEULAND Innovation Contest, KIT Research to Business Online Portal (https://www.kit-technology.de/en/blog/interview-nominees-of-the-neuland-innovation-contest, 26.04.2021, 16:10)

BattMarines

Abdalla Batta, Andrea Bellelli, Karsten Litfin, Andreas Class

A mayor challenge of the Energiewende is energy storage to bridge periods where neither wind nor photovoltaics can provide renewable energy sources. For the success of storage solutions in the market, both a low price and an acceptable efficiency are equally relevant factors. Power-Heat-Power solutions for energy storage accept compromises with respect to the achievable efficiency and aim at low system prices to become competitive. The novelty of the proposed BattMarines is to use a conventional Rankine-cycle in batch operation as an energy storage system and to exploit scaling to large dimensions (storage capacity ~1 GWd) to reach economical goals.

Figure 5 shows a schematic sketch of BattMarines [1]. A high-efficiency steam turbine power plant is installed on a floating platform on the sea. A huge tank resides on the seabed and contains large quantities of hot live steam. To produce electric power on demand steam is drawn through the vertical steam line connecting the platform and the storage vessel. Charging the system is accomplished by electrically heating fresh water when a surplus of renewable energy becomes available. Even though the concept appears to be naïve, it offers remarkable advantages with respect to its scalability to the anticipated large storage capacity. The steam within the thin-shell storage vessel is stored at the ambient pressure of the surrounding water at large depth and thus exhibits high energy density. The heat losses of the storage vessel and the steam line are obviously quiet large but in comparison to the huge energy contents, they become acceptable, so that storage times ranging from several hours to seasonal storage become feasible.

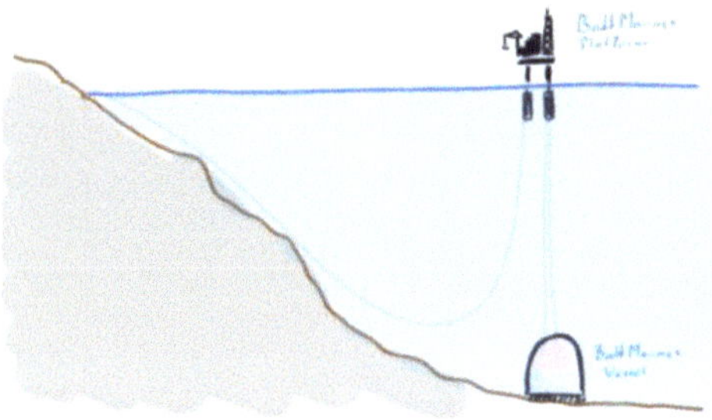

Figure 5 Sketch of BattMarines Power-Heat-Power storage concept.

References

[1] Class, A. (2020); BattMarines, 2nd International Workshop on Carnot Batteries 2020, University Stuttgart, September 15.-16, 2020.

List of Publications

[1] Bai, S. Agent based model in risk management: several applications. The Second International Symposium on Simulation and Process Modelling (ISSPM 2020), Shenyang, China, 29.–30. August 2020.

[2] Bai, S. Simulations of COVID-19 spread by spatial agent-based model and ordinary differential equations. International journal of simulation and process modeling [modelling], 15 (3), 268–277. doi:10.1504/IJSPM.2020.10029411

[3] Bai, S.; Raskob, W. Agent-based Negotiation Simulation. 5th NERIS Workshop "Key challenges in the preparedness, response and recovery phase of a nuclear or radiological emergency". Proceedings, 53–62.

[4] Bai, S.; Raskob, W.; Müller, T. Agent based model. Radioprotection, 55 (HS1), S187–S191. doi:10.1051/radiopro/2020031

[5] Batta, A.; Class, A. Thermal Hydraulic CFD Validation for Liquid Metal Cooled 19-Pin Hexagonal Wire Wrapped Rod Bundle. 18th International Topical Meeting on Nuclear Reactor Thermal Hydraulics 2019, (NURETH 2019), Portland, Oregon, USA, August 18-23, 2019. Vol. 1, 655–665, Curran Associates, Inc.

[6] Bolleman, J. T.; Yokoyama, T. T.; Heumos, S.; Seaman, J.; Trybushnyi, D.; Pook, T.; Guarracino, A.; Garrison, E. (2020, Juli 16). Semantic Variation Graphs: Ontologies for Pangenome Graphs. International Conference on Intelligent Systems for Molecular Biology (2020), Online, 13.–16. Juli 2020.

[7] Bühler, L.; Biao, L.; Brinkmann, H.-J.; Mistrangelo, C. Reconstruction of 3D fluid velociy and electric potential from measurements on the external surface of a thick-walled pipe. 31st Symposium on Fusion Technology (SOFT 2020), Online, 20.–25. September 2020.

[8] Bühler, L.; Brinkmann, H.-J.; Mistrangelo, C. Experimental investigation of liquid metal pipe flow in a strong non-uniform magnetic field. Magnetohydrodynamics, 56 (2-3), 131–138. doi:10.22364/mhd.56.2-3.4

[9] Bühler, L.; Mistrangelo, C.; Brinkmann, H.-J. Experimental investigation of liquid metal MHD flow entering a flow channel insert. Fusion engineering and design, 154, Article No.111484. doi:10.1016/j.fusengdes.2020.111484

[10] Class, A. G.; Gabriel, S. BattMarines. 90th Executive Committee Meeting (XC90) of the Energy Storage IEA Technology Collaboration Platform (2020), Online, 18.–19. November 2020.

[11] Class, A. G.; Gabriel, S. BattMarines. 2nd International Workshop on Carnot Batteries (2020), Stuttgart, Deutschland, 15.–16. September 2020.

[12] Dietrich, B. NECOC macht aus CO2 einen Hightech-Rohstoff. Klima und Umwelt News: Newsletter des KIT-Zentrums Klima und Umwelt, 2, 1.

[13] Dietrich, B. Vom Treibhausgas zum Hightech-Rohstoff – BMWi-Forschungsprojekt NECOC. GWF / Gas, Erdgas, 11-12, 1–9.

[14] Duranova, T.; Bedwell, P.; Beresford, N. A.; Bleher, M.; Gering, F.; Geertsema, G.; Hamburger, T.; Kaiser, J. C.; Korsakissok, I.; Leadbetter, S. J.; Montero, M.; Müller, T.; Oughton, D.; Perko, T.; Raskob, W.; Tomas, J.; Turcanu, C.; de Vries, H.; Walsh, L.; Woda, C. CONFIDENCE dissemination meeting: Summary on the scenario-based workshop. Radioprotection, 55 (HS1), S17–S37. doi:10.1051/radiopro/2020009

[15] Duranova, T.; Beresford, N. A.; Perko, T.; Raskob, W. Education and training activities in the Euratom CONFIDENCE project. Radioprotection, 55 (HS1), S45–S50. doi:10.1051/radiopro/2020011

[16] Duranova, T.; van Asselt, E.; Müller, T.; Bohunova, J.; Twenhöfel, C. J. W.; Smetsers, R. C. G. M. MCDA stakeholder workshops. Radioprotection, 55 (HS1), S193–S196. doi:10.1051/radiopro/2020032

[17] Flesch, J.; Marocco, L.; Fritsch, A.; Niedermeier, K.; Wetzel, T. Entropy generation minimization analysis of solar salt, sodium, and lead–bismuth eutectic as high temperature heat transfer fluids. Journal of heat transfer, 142 (4), Article: 042103. doi:10.1115/1.4046302

[18] Gaus-Liu, X.; Cron, T.; Fluhrer, B. Experimental Studies on Two-Layer Corium Heat Transfer in Light Water Reactor Lower Head in LIVE2D Facility. Nuclear technology, 206 (9), 1385–1396. doi:10.1080/00295450.2020.1743102

[19] Gaus-Liu, X.; Cron, T.; Fluhrer, B. LIVE2D Experiments on Thermohydraulics of Stratified Melt Pool in LWR Lower Plenum. International Seminar "In-vessel retention: outcomes of IVMR project", Juan les Pins, France, January 21-22, 2020.

[20] Gaus-Liu, X.; Cron, T.; Fluhrer, B.; Komlev, A.; Ma, W.; Bechta, S. Experiments on Thermohydraulics of Stratified Melt Pool – SIMECO-II and LIVE2D Tests. International Seminar "In-vessel retention: outcomes of IVMR project" (2020), Juan les Pins, France, 21.–22. Januar 2020.

[21] Gianfelici, S.; Perez-Martin, S.; Pfrang, W.; Schulenberg, T.; Rineiski, A. Validation of the FEMAXI/SIMMER coupling by using the E9 CABRI-2 experiment. PHYSOR 2020: Transition to a Scalable Nuclear Future, Cambridge, United Kingdom, March 29th. doi:10.1051/epjconf/202124707015

[22] Hu, G.; Zhang, H.; Liu, Q. Review on sensors to measure control rod position for nuclear reactor. Annals of nuclear energy, 144, Article No.107485. doi:10.1016/j.anucene.2020.107485

[23] Jordan, T. (2020, Dezember 14). PRESLHY Experiment series E5.1 (Ignited Discharge) results – Benchmarking. doi:10.5445/IR/1000127300

[24] Kaiser, F.; Gabriel, S. COSMOS-L: Flow during Critical Heat Flux. doi:10.5445/IR/1000127105

[25] Kaiser, F.; Gabriel, S. COSMOS-L: Forced convection boiling in a concentric annular gap - T80G400P1200 - 75% of CHF power. doi:10.5445/IR/1000127047

[26] Kaiser, F.; Gabriel, S. COSMOS-L: Forced convection boiling in a concentric annular gap - T80G600P2000 - 50% of CHF power. doi:10.5445/IR/1000127048

[27] Kaiser, F.; Gabriel, S. COSMOS-L: Forced convection boiling in a concentric annular gap - T80G400P1200 - 50% of CHF power. doi:10.5445/IR/1000127046

[28] Kaiser, F.; Gabriel, S. COSMOS-L: Forced convection boiling in a concentric annular gap - T80G600P2000 - 90% of CHF power. doi:10.5445/IR/1000127044

[29] Kaiser, F.; Gabriel, S. COSMOS-L: Forced convection boiling in a concentric annular gap - T80G600P2000 - 75% of CHF power. doi:10.5445/IR/1000127045

[30] Kaiser, F.; Gabriel, S. COSMOS-L: Forced convection boiling in a concentric annular gap - T80G600P2000: macroscopic view. doi:10.5445/IR/1000127043

[31] Kaiser, F.; Gabriel, S. COSMOS-L: Forced convection boiling in a concentric annular gap - T80G400P2000: macroscopic view. doi:10.5445/IR/1000126774

[32] Kaiser, F.; Gabriel, S. COSMOS-L: Forced convection boiling in a concentric annular gap - T80G400P1200: macroscopic view. doi:10.5445/IR/1000126773

[33] Kaiser, F.; Gabriel, S. COSMOS-L: Forced convection boiling in a concentric annular gap - T65G400P2000 - 90% of CHF power. doi:10.5445/IR/1000126772

[34] Kaiser, F.; Gabriel, S. COSMOS-L: Forced convection boiling in a concentric annular gap - T65G400P2000: macroscopic view. doi:10.5445/IR/1000126771

[35] Kaiser, F.; Gabriel, S. COSMOS-L Forced Convection boiling in a concentric annular gap - T80G400P1200 - 90% of CHF power. doi:10.5445/IR/1000126560

[36] Kaiser, F.; Gabriel, S. COSMOS-L Forced Convection boiling in a concentric annular gap - T80G400P2000 - 50% of CHF power. doi:10.5445/IR/1000126559

[37] Kaiser, F.; Gabriel, S. COSMOS-L Forced Convection boiling in a concentric annular gap - T80G400P2000 - 75% of CHF power. doi:10.5445/IR/1000126393

[38] Kaiser, F.; Gabriel, S. COSMOS-L Forced Convection boiling in a concentric annular gap - T80G400P2000 - 90% of CHF power. doi:10.5445/IR/1000125546

[39] Kamenik, B.; Gaus-Liu, X.; Marn, J.; Kljenak, I. Simulation of LIVE2D Experiment on Reactor Core Melt in Reactor PressureVessel Lower Plenum. 29th International Conference Nuclear Energy for New Europe: September 9-10, Portoroz, Slovenia: NENE 2020 Conference Proceedings (NENE 2020). Ed.: L. Snoj, 1804.1–1804.8.

[40] Kamienowska, M.; Stoppel, L.; Niedermeier, K.; Bender, M.; Wetzel, T. Synthesis of anhydrous formaldehyde. Jahrestreffen der ProcessNet-Fachgruppe Wärme- und Stoffübertragung (2020), Erfurt, Deutschland, 12.–13. März 2020.

[41] Klüber, V.; Bühler, L.; Mistrangelo, C. Numerical simulation of 3D magnetohydrodynamic liquid metal flow in a spatially varying solenoidal magnetic field. Fusion engineering and design, 156, Art.Nr. 111659. doi:10.1016/j.fusengdes.2020.111659

[42] Laube, T.; Marocco, L.; Niedermeier, K.; Pacio, J.; Wetzel, T. Thermodynamic Analysis of High-Temperature Energy Storage Concepts Based on Liquid Metal Technology. Energy technology, 8 (3), Article no: 1900908. doi:10.1002/ente.201900908

[43] Leung, L. K. H.; Guzonas, D.; Schulenberg, T. Special Section: Selected Papers From the Ninth International Symposium On Supercritical Water-cooled Reactors. Journal of nuclear engineering and radiation science, 6 (3), Art.-Nr. 030301. doi:10.1115/1.4046827

[44] Li, Y.; Xiao, J.; Zhang, H.; Breitung, W.; Travis, J.; Kuznetsov, M.; Jordan, T. Numerical analysis of hydrogen release, dispersion and combustion in a tunnel with fuel cell vehicles using all-speed CFD code GASFLOW-MPI. International journal of hydrogen energy. doi:10.1016/j.ijhydene.2020.09.063

[45] Li, Y.; Zhang, H.; Xiao, J.; Peter, R.; Jordan, T. Numerical simulation of the long-term radiolytic gas accumulation in diversified measurement pipeline for BWR using GASFLOW-MPI. Progress in nuclear energy, 123, Art.-Nr.: 103319. doi:10.1016/j.pnucene.2020.103319

[46] Marocco, L.; Dietrich, B.; Wetzel, T.; E-ckel, A.-M. LES of turbulent aided mixed convection to a low Prandtl number fluid in a concentric annulus. 38th International Conference on Heat and Mass Transfer (2020), Gaeta, Italien, 24.–26. Juni 2020.

[47] Massone, M.; Wang, S.; Rineiski, A.; Servell, P. Dimensioning of the emergency draining tank for a molten salt reactor through analytical modeling. Annals of nuclear energy, 138, Article No.107121. doi:10.1016/j.anucene.2019.107121

[48] Mistrangelo, C.; Bühler, L.; Klüber, V. Three-dimensional magneto convective flows

in geometries relevant for DCLL blankets. Fusion engineering and design, 159, Article: 111686. doi:10.1016/j.fusengdes.2020.111686

[49] Mistrangelo, C.; Bühler, L.; Klüber, V.; Smolentsev, S.; Maione, I.; Aubert, J. MHD flow in liquid metal blankets: major design issues, numerical analysis and code validation. 31st Symposium on Fusion Technology (SOFT 2020), Online, 20.–25. September 2020.

[50] Möhrle, S. (2020, Juni 16). Case-Based Decision Support for Disaster Management. Dissertation. KIT Scientific Publishing. doi:10.5445/KSP/1000099369

[51] Müller, T.; Bai, S.; Raskob, W. MCDA handling uncertainties. Radioprotection, 55 (HS1), S181–S185. doi:10.1051/radiopro/2020030

[52] Müller-Trefzer, F. Hochtemperaturwärmespeicher mit Flüssigmetall-Dual-Media Schichtenspeicher. Symposium "Energie-Campus" der Stiftung Energie & Klimaschutz im Rahmen des Ideenwettbewerbs "Energie und Umwelt - Meine Idee für morgen" (2020), Stuttgart, Deutschland, 20. November 2020.

[53] Nagy, A.; Perko, T.; Müller, T.; Raskob, W.; Benighaus, L. Uncertainty visualization using maps for nuclear and radiological emergencies. Radioprotection, 55 (HS1), S197–S203. doi:10.1051/radiopro/2020033

[54] Niedermeier, K.; Müller-Trefzer, F.; Daubner, M.; Marocco, L.; Weisenburger, A.; Wetzel, T. Theoretical and Experimental Studies of Dual-Media Thermal Energy Storage with Liquid Metal. 26th SolarPACES (2020), Online, 28. September–2. Oktober 2020.

[55] Niedermeier, K.; Müller-Trefzer, F.; Weisenburger, A.; Wetzel, T. High-temperature storage with liquid metals - Design of a prototype storage system and material testing. 2nd International Workshop on Carnot Batteries (2020), Stuttgart, Deutschland, 15.–16. September 2020.

[56] Ottenburger, S. S.; Bai, S.; Raskob, W. MCDA-based Genetic Algorithms for Developing Disaster Resilient Designs of Critical Supply Networks. 2019 International Conference on Information and Communication Technologies for Disaster Management (ICT-DM), Article No.9032982, Institute of Electrical and Electronics Engineers (IEEE). doi:10.1109/ICT-DM47966.2019.9032982

[57] Ottenburger, S. S.; Kemal Çakmak, H.; Jakob, W.; Blattmann, A.; Trybushnyi, D.; Raskob, W.; Kühnapfel, U.; Hagenmeyer, V. A Novel Optimization Method for Urban Resilient and Fair Power Distribution Preventing Critical Network States. International journal of critical infrastructure protection, 29, Art.Nr. 100354. doi:10.1016/j.ijcip.2020.100354

[58] Otten, C.; Schulz, B.; Teitz, S.; Eichinger, F.; Seibt, A.; Kuhn, D.; Würdemann, H. Interactions between a calcium scaling inhibitor, geothermal fluids, and microorganisms - Results of in situ monitoring in the molasse basin and laboratory experiments. EGU General Assembly - Sharing Geoscience Online (EGU 2020), Wien, Österreich, 4.–8. Mai 2020. doi:10.5194/egusphere-egu2020-22434

[59] Raskob, W.; Beresford, N. A.; Duranova, T.; Korsakissok, I.; Mathieu, A.; Montero, M.; Müller, T.; Turcanu, C.; Woda, C. Confidence: achievements and way forward. Radioprotection, 55 (HS1), S39–S43. doi:10.1051/radiopro/2020010

[60] Raskob, W.; Beresford, N. A.; Duranova, T.; Korsakissok, I.; Mathieu, A.; Montero, M.; Müller, T.; Turcanu, C.; Woda, C. CONFIDENCE: project description and main results. Radioprotection, 55 (HS1), S7–S15. doi:10.1051/radiopro/2020008

[61] Raskob, W.; Duranova, T. Editorial: the main results of the European CONFIDENCE project. Radioprotection, 55 (HS1), S1–S5. doi:10.1051/radiopro/2020007

[62] Raskob, W.; Haller, C.; Hasemann, I.; Schichtel, T.; Trybushnyi, D. Source term reconstruction module in JRODOS. 5th NERIS Workshop "Key challenges in the preparedness, response and recovery phase of a nuclear or radiological emergency". Proceedings, 112–118.

[63] Reimann, J.; Vicente, J.; Ferrero, C.; Rack, A.; Gan, Y. 3d tomography analysis of the packing structure of spherical particles in slender prismatic containers. International journal of materials research, 111 (1), 65–77. doi:10.3139/146.111814

[64] Schulenberg, T. Die vierte Generation der Kernreaktoren – Grundlagen, Typen und Nutzen verständlich erklärt. Springer Verlag. doi:10.1007/978-3-662-61605-5

[65] Schulenberg, T.; Heinze, D. Numerical Simulation of a Steam Injector for Passive Residual Heat Removal. Journal of nuclear engineering and radiation science, 6 (3), 031105. doi:10.1115/1.4045692

[66] Sheikh, M. A. R.; Liu, X.; Matsumoto, T.; Morita, K.; Guo, L.; Suzuki, T.; Kamiyama, K. Numerical Simulation of the Solid Particle Sedimentation and Bed Formation Behaviors Using a Hybrid Method. Energies, 13 (19), Art.-Nr. 5018. doi:10.3390/en13195018

[67] Smolentsev, S.; Rhodes, T.; Yan, Y.; Tassone, A.; Mistrangelo, C.; Bühler, L.; Urgorri, F. R. Code-to-Code Comparison for a PbLi Mixed-Convection MHD Flow. Fusion science and technology, 76 (5), 653–669. doi:10.1080/15361055.2020.1751378

[68] Stokmaier, M. J. (2020, Mai 4). Design and optimisation of acoustic resonators for sonofusion experiments (Design und Optimierung akustischer Resonatoren für Sonofusionsexperimente). Dissertation. Karlsruher Institut für Technologie (KIT). doi:10.5445/IR/1000118780

[69] Tarantino, M.; Roelofs, F.; Shams, A.; Batta, A.; Moreau, V.; Di Piazza, I.; Gershenfeld, A.; Planquart, P. SESAME project: advancements in liquid metal thermal hydraulics experiments and simulations. EPJ Nuclear Sciences & Technologies, 6 (18). doi:10.1051/epjn/2019046

[70] Tian, W.; Zhang, H.; Liu, Q.; Hu, G.; He, W. The experimental study of single droplet impinging on an inclined heated wall. 2020 International Conference on Nuclear Engineering, ICONE 2020, collocated with the ASME 2020 Power Conference, 4-5 April 2020, online, The American Society of Mechanical Engineers (ASME).

[71] Veiga-López, F.; Kuznetsov, M.; Martínez-Ruiz, D.; Fernández-Tarrazo, E.; Grune, J.; Sánchez-Sanz, M. Unexpected Propagation of Ultra-Lean Hydrogen Flames in Narrow Gaps. Physical review letters, 124 (17), Art.-Nr.: 174501. doi:10.1103/PhysRevLett.124.174501

[72] Veiga-López, F.; Martínez-Ruiz, D.; Kuznetsov, M.; Sánchez-Sanz, M. Thermoacoustic analysis of lean premixed hydrogen flames in narrow vertical channels. Fuel, 278, Art.-Nr. 118212. doi:10.1016/j.fuel.2020.118212

[73] Vogel, J.; Keller, M.; Johnson, M. Numerical modeling of large-scale finned tube latent thermal energy storage systems. Journal of energy storage, 29, Art.-Nr.: 101389. doi:10.1016/j.est.2020.101389

[74] Walsh, L.; Ulanowski, A.; Kaiser, J. C.; Woda, C.; Raskob, W. A new European cancer risk assessment tool for application after nuclear accidents. Radioprotection, 55 (HS1), S95–S99. doi:10.1051/radiopro/2020018

[75] Wandler, S.; Bai, S.; Raskob, W. Energy consumption models in dialysis clinics for agent-based decision support. International journal of simulation and process modeling [modelling], 15 (1-2), 170–188. doi:10.1504/IJSPM.2020.106981

[76] Wang, Z.; Pan, X.; Jiang, Y.; Wang, Q.; Li, Y.; Xiao, J.; Jordan, T.; Jiang, J. Experimental study on shock wave propagation and spontaneous ignition induced by high-pressure hydrogen suddenly released into T-shaped tubes. Safety science, 127, Art.-Nr. 104694. doi:10.1016/j.ssci.2020.104694

[77] Wang, Z.; Pan, X.; Jiang, Y.; Wang, Q.; Yan, W.; Xiao, J.; Jordan, T.; Jiang, J. Experiment study on the pressure and flame characteristics induced by high-pressure hydrogen spontaneous ignition. International journal of hydrogen energy, 45 (35), 18042–18056. doi:10.1016/j.ijhydene.2020.04.051

[78] Wang, Z.; Zhang, H.; Pan, X.; Jiang, Y.; Wang, Q.; Xiao, J.; Jordan, T.; Jiang, J. Experimental and numerical study on the high-pressure hydrogen jet and explosion induced by sudden released into the air through tubes. International journal of hydrogen energy, 45 (7), 5086–5097. doi:10.1016/j.ijhydene.2019.12.072

[79] Yokoyama, T. T.; Heumos, S.; Seaman, J.; Trybushnyi, D.; Pook, T.; Guarracino, A.; Garrison, E.; Bolleman, J. T. (2020, Juli 15). Semantic Variation Graphs: Ontologies for Pangenome Graphs. International Conference on Intelligent Systems for Molecular Biology (2020), Online, 13.–16. Juli 2020.

[80] Yu, F. (2020, August 10). Numerical Studies of Nuclear Containment Spray Process by Stochastic Field Method and CGCFD approach. Dissertation. Karlsruher Institut für Technologie (KIT). doi:10.5445/IR/1000122409

[81] Yu, F.; Ji, Y.; Zhang, H.; Class, A.; Xiao, J.; Jordan, T. Development of reverse meshing technology on geometry reconstruction for CFD code GASFLOW-MPI. Progress in nuclear energy, 118, Art.-Nr.: 103124. doi:10.1016/j.pnucene.2019.103124

[82] Zhang, H.; Ma, Y.; Hu, G.; Liu, Q. Droplet impaction in nuclear installations and safety analysis: Phenomena, findings and approaches. Nuclear engineering and design, 366, Art.-Nr. 110757. doi:10.1016/j.nucengdes.2020.110757